Practical Radio Frequency Test and Measurement:
A Technician's Handbook

Practical Radio Frequency Test and Measurement: A Technician's Handbook

Joseph J. Carr

Newnes

An Imprint of Elsevier

Amsterdam   Boston   Heidelberg   London   New York   Oxford
Paris   San Diego   San Francisco   Singapore   Sydney   Tokyo

Newnes is an imprint of Elsevier.

Copyright © 2002, Elsevier.

Carr, Joseph J.
    Practical radio frequency test and measurement : a technician's handbook /
        Joseph J. Carr

    p.   cm.
    Includes bibliographical references and index.
    ISBN-13: 978-0-7506-7161-3   ISBN-10: 0-7506-7161-0 (pbk.: alk. paper)
    1. Radio measurements.   2. Electronic measurements.   3. Radio circuits - Testing.
I. Title.
TK6553.C2943   1999   99-25672
621.384-dc21   CIP
ISBN-13: 978-0-7506-7161-3
ISBN-10: 0-7506-7161-0
British Library Cataloguing-in-Publication Data
A catalogue record for this book is available from the British Library.

The publisher offers special discounts on bulk orders of this book.
For information, please contact:
Manager of Special Sales
Elsevier
200 Wheeler Road
Burlington, MA 01803
Tel: 781-313-4700
Fax: 781-313-4802

For information on all Newnes publications available, contact our World
Wide Web homepage at http://www.newnespress.com

10 9 8 7 6

Printed in the United States of America

This book is dedicated to the fond memory of
Michael J. Shaffer, D.Sc.
Dr. Shaffer was Director, Bioelectronics Laboratory and
Professor of Anesthesiology (Emeritus),
George Washington University Medical Center
Apollo Space Capsule Life Support System
Flight Director, Biosatellite II
Radar and Telecommunications Engineer
Signals Officer, Royal Signal Corps
(British Army World War II)
Engineer, Mentor, Boss, and Friend

CONTENTS

PREFACE

This book is about making measurements on radio frequency devices and in radio frequency systems. "RF" is different from low frequencies for a number of reasons. The chief reason is that component sizes approximate wavelengths in many cases, so they become very important. Also, in certain cases stray inductance and capacitance have a much more profound effect on RF circuits than they do on lower frequency circuits. In this book you will find information about a number of topics, including the basic theories of RF and any other form of measurement, as well as some useful RF components for test and measurement. The Smith Chart is covered (it would be a contradiction to leave it out). Additional topics include: signal sources and signal generators; spectrum and network analyzers; RF power measurement; measurement of time, frequency, and period; radio receiver measurements; radio transmitter measurements; RF amplifier (including high-power amplifier) measurements; antenna gain and pattern measurement; antenna and transmission line measurements; L-C-R measurements at RF frequencies; and manual Time Domain Reflectometry (TDR) measurements.

My approach taken is to examine the instruments used in the various types of measurement and then to look at practical measurement methods. The latter is not always easily understood without knowledge of the former.

Joseph J. Carr, MSEE
Falls Church, Virginia

CHAPTER ONE

Introduction to Radio Frequency Electronics and Measurement Theory

The physicist Lord Kelvin once remarked that one doesn't really understand a thing until one can measure it and assign numbers to it. Measurement is the assignment of numbers to represent the attributes or properties of something, according to a standard rule that gives the numbers an ordered meaning. Measurements are also used for qualitative data, but the results tend to be bivalent (two-valued), such as true or false, yes or no, 1 or 0, and so on. This book discusses the various methods used for measuring radio frequency circuits and systems. But first I will discuss the basic theories of all forms of electronic measurements, which will help the rest of the material make more sense.

MEASUREMENTS

Measurements are made to fulfill any or all of several different goals:

- Obtain information about a physical phenomenon;
- Assign a value to some fundamental constant;
- Record trends in, or control, some process; and
- Correlate behavior of a phenomenon with other parameters in order to obtain insight into their relationships.

Measurement is an act designed to "derive quantitative information about" some phenomenon "by comparison to a reference" (Herceg, 1972). The physical quantity being measured is called the *measurand* or *factor*.

Measurement Data Classes

The data that results from measurements can be divided into two major classes, and those classes are each divided into two subclasses. The major divisions include *qualitative* data and *quantitative* data.

Qualitative Data

This type of data is nonnumerical or categorical in nature. It includes things like the presence or absence of some factor, good or bad, defective or not defective, gender, race, and so forth.

Qualitative data doesn't inherently result in numbers, so it is sometimes held in less esteem than quantitative data. This attitude is mistaken, unless there is some inherent need for numbers in a particular case.

Qualitative data can be further broken into two subgroups: *nominal* data and *ordinal* data.

Nominal data is qualitative data that has no inherent ordering or rank. Examples include lists of names, labels, groupings, and so forth. *Ordinal* data allows ranking, but differences between data points are either nonexistent or meaningless.

Qualitative data can sometimes be given a numerical flavor by correct collection techniques. For example, because much of this data is bivalent in nature, one can assign the digits 0 and 1 (e.g., 1 for "yes" and 0 for "no"). Another popular method is to assign some arbitrary but consistent scale indicating depth of feeling, preferences, and so forth. For example, a scale is often used to assign numerals to questions that are largely qualitative value judgments.

Quantitative Data

Quantitative data is that which naturally results in some number to represent the factor. Examples include amount of money, length, temperature, number of defects per unit, voltage, pressure, weight, frequency, period, phase, power, and so on.

Quantitative data can be further divided into two subclasses: *interval* data and *ratio* data. *Interval data* allows for a meaningful comparison of differences, but not the relative values of two or more factors. Such measurements are made relative to an arbitrarily selected standard zero point. For example, in the West we assign calendar dates according to the supposed birth date of Christ. In other cultures, the zero point is some other fixed historical event.

Another example of an interval measure is temperature, as measured on the Celsius (centigrade) and Fahrenheit temperature scales (°C and °F, respectively). The Celsius scale sets the zero degree point at the freezing point of water. Another arbitrary (but convenient) reference point on the Celsius scale is the 100 °C point: the boiling point of water.

Why are these points used to define the Celsius scale? Aren't they somewhat arbitrary? The reason these points are used is that they are easy to replicate whenever anyone wants to calibrate a Celsius thermometer. In other words, one doesn't need a high-quality metrology laboratory to do a reasonably good job of calibrating a common thermometer.

The Celsius scale is also sometimes called "centigrade" because there are 100 (*centi*) equal divisions between the arbitrarily set 0 °C and 100 °C points. The zero point on the Fahrenheit scale is equally arbitrary, but its selection seems a bit irrational (water freezes at 32 °F).

Why are the 0 °C and 100 °C points on the Celsius scale arbitrary? Because there is no particular compelling reason to select these points, except for calibration convenience. After all, there are temperatures colder than 0 °C and hotter than 100 °C.

The selection of zero points on the temperature scale illustrates the properties of interval data: we can make meaningful statements about differences of temperature, but differences cannot be scaled up; that is, 40 °C is not twice as much temperature as 20 °C (see Figure 1.1).

Ratio data are based on some fixed or natural zero point: for example, weights, pressures, and temperatures such as the Kelvin scale. The Kelvin temperature scale uses degrees of the same size as in the Celsius scale (a change of 1 °C is the same as a change of 1 °K), but the zero reference point is what physicists call "Absolute zero"—that is, the temperature at which all molecular motion ceases (0 °K is about −273.16 °C). Thus, 0 °K represents a natural zero point.

A consequence of having a natural zero reference point is that ratios as well as differences are meaningful. Raising a temperature from 100 °K to 200 °K is an increase of twice the temperature.

Measurement Standards

Metrology, the science of measurement, requires a "rule" to which things are compared; that rule is called a *standard.* Not all standards are equal, so there is a hierarchy of standards (Figure 1.2): *international reference standards, primary standards, transfer standards, working standards,* and *shop-level standards.*

International Reference Standards

These standards are those agreed upon by an international standards organization. For years, the reference standard for the meter was a platinum bar, 1.0000 meters long, stored in a vault in Paris, France, maintained by the *International Standards Institute.* Various authorities around the world keep other international standards.

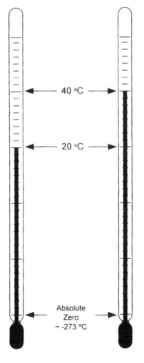

Figure 1.1
When examined against absolute zero, 40 °C is not twice 20 °C, despite what the relative numbers might indicate.

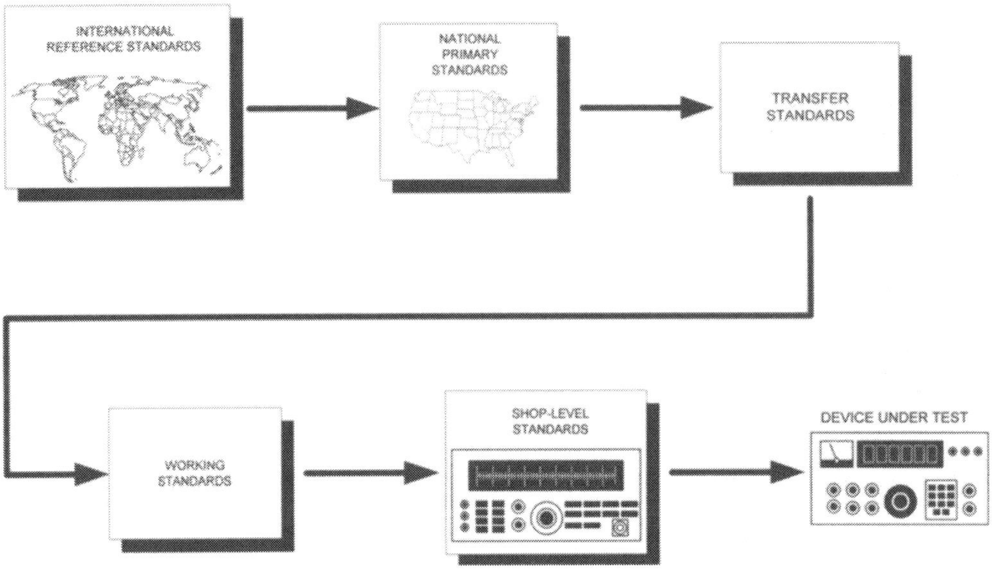

Figure 1.2   Hierarchy of standards for testing instruments.

Primary Standards

These are the principal standards maintained at a national level in the various countries. In the United States, these standards are maintained by the *National Institutes of Standards and Technology* (NIST), formerly, named the *National Bureau of Standards* (NBS). Some primary standards are periodically compared with an international reference standard maintained by a group such as the International Standards Institute.

Transfer Standards

These standards are second level, and are periodically compared with the primary standard. They are used to calibrate lower-order standards used in the country so that wear and tear on the primary standard is reduced.

Working Standards

The working standard is compared with the transfer standard in a nationally certified laboratory, or at NIST. Such standards are said to be "NIST traceable."

Shop Level and Secondary Standards

These standards are used locally (e.g., in the lab) to calibrate instruments and gauges.

Gauges/Instruments

The lowest order of standards, these are the devices actually used to make measurements and collect data on the objects being measured. Gauges and Instruments are compared with either working standards or secondary standards.

Categories of Measurement

There are three general categories of measurement: *direct, indirect,* and *null.* Electronic instruments are available based on all three categories.

Direct measurements are made by holding the measurand up to some calibrated standard and comparing the two. A good example is the ruler used to cut a piece of coaxial cable to the correct length. You know that the cable must be cut to a length of 56 cm, so you hold a meter stick (the standard or reference) up to the uncut piece of cable (Figure 1.3). Set the "0 cm" point at one end, and make a mark on the cable adjacent to the "56" mark on the meter stick, and then make your cut at the appropriate point.

Indirect measurements are made by measuring something other than the actual measurand. Although frequently considered "second best" from the perspective of measurement accuracy, indirect methods are often used when direct measurements are either difficult or dangerous. For example, one might measure the temperature of a point on the wall of a furnace that is melting metal, knowing that it is related to the interior temperature by a certain factor. There was once a minicomputer manufacturer who used an indirect temperature measurement to ease the job of the service technicians. The manufacturer drilled a small hole at the top of the rack-mounted cabinet where the temperature would be <39 °C when the temperature on the electronic circuit boards deep inside the cabinet was within specification. Although the technicians were interested in the temperature at the board level, they actually make a measurement that correlates to the desired measurement.

The system manufacturer specified this method for two reasons:

1. The measurement point was easily available (while the boards were not) and thus did not require any disassembly.
2. The service technician could use an ordinary medical fever thermometer (30 °C to 42 °C) from a corner drug store as the measurement instrument.

An example of indirect measurement in radio frequency (RF) is measuring the heat generated when RF power is dissipated in a resistive load, and then calculating the RF power level from the temperature change information.

Null measurements are made by comparing a known (calibrated) value to an unknown measurand, and then adjusting either one or the other until the difference between them is zero. An *electrical potentiometer* is such an instrument; it is an

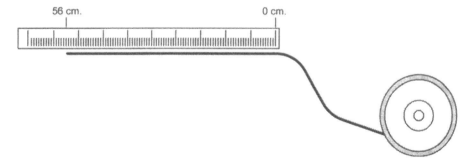

Figure 1.3   Measuring cable with a ruler.

adjustable calibrated voltage source and a comparison meter (zero-center gal-vanometer). The reference voltage from the potentiometer is applied (Figure 1.4) to one side of the zero-center galvanometer (or one input of a difference measuring voltmeter), and the unknown is applied to the other side of the galvanometer (or remaining input of the differential voltmeter). The output of the potentiometer is adjusted until the meter reads zero difference. The setting of the potentiometer under the null condition is the same as the unknown measurand voltage.

Quality Factors in Making Measurements

The "goodness" of electronic measurements involves several concepts that must be understood. Some of the more significant of these are *error, validity, reliability, repeatability, accuracy, precision,* and *resolution.*

Error

In all measurements there is a certain degree of error present. The word "error" in this context refers to normal, random variations and in no way means "mistakes." In short order we will discuss error in greater depth.

If measurements are made repeatedly on the same parameter (one which is truly unchanging), or if different instruments or instrument operators are used to make successive measurements, it will be found that the measurements tend to cluster around a central value (X_o in Figure 1.5). In most cases, it is assumed that X_o is the true value, but if there is substantial inherent error in the measurement process, then it may deviate from the true value (X_i) by a certain amount (ΔX)—which is the error term. The assumption that the central value of a series of measurements is the true value is only valid when the error term is small, as in $\Delta X \rightarrow 0$, $X_o \rightarrow X_i$.

Validity

The *validity* of a measurement is a statement of how well the instrument actually measures what it purports to measure. What determines the validity of a sensor

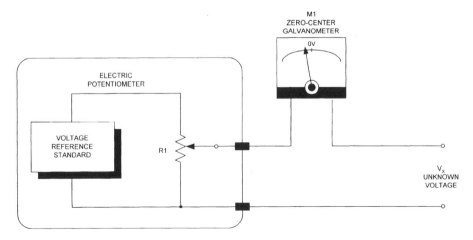

Figure 1.4 Potentiometric comparison measurement.

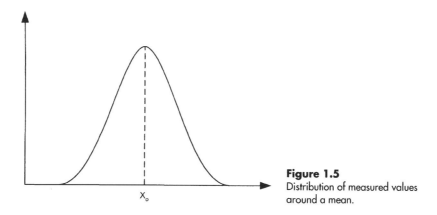

Figure 1.5
Distribution of measured values
around a mean.

measurement is the extent to which the actual measurement relates to the desired measurement, and over what range or under what conditions. Many measurement devices exist where the output readings are only meaningful under certain specified conditions or over a specified range. A diode sensor used to measure RF power has a specified range of power levels for which it is valid.

Reliability and Repeatability

The *reliability* of the measurement is a statement of its *consistency* when discerning the values of the measurand on different trials, when the measurand may take on very different values. For example, a thermistor, which is used in an RF wattmeter to measure the heat dissipated in a resistive load by RF power, may have its resistance value and resistance change curve altered by overheating.

Related to reliability is the idea of *repeatability,* which refers to the ability of the instrument to return the same value when repeatedly exposed to the exact same stimulant. Neither reliability nor repeatability is the same as accuracy, for a measurement may be both "reliable" and "repeatable" while being quite wrong.

Accuracy and Precision

The *accuracy* of a measurement refers to the freedom from error, or the degree of conformance between the measurand and the standard. *Precision,* on the other hand, refers to the exactness of successive measurements, also sometimes considered the degree of refinement of the measurement. Accuracy and precision are often confused with one another, and these words are often erroneously used interchangeably. One way of stating the situation is to note that *a precise measurement has a small standard deviation and variance under repeated trials,* but *in an accurate measurement the mean value of the normal distribution curve is close to the true value.*

The relationship between precision and accuracy can be seen in the target-shooting example of Figure 1.6. In all of these cases, the data form a normal distribution curve when repeatedly performed over a large number of iterations of the measurement. Four targets are shown in a precision-versus-accuracy matrix. Target 1-6A has *good accuracy* because the shots are clustered on the bull's eye. It also has *good precision,* as seen by the fact that the cluster has a small dispersion—that is, it is a "tight group," as target shooters say. The target at 1-6B has good precision (small dispersion, good clustering), but the cluster is off-center, high, and to the left.

The target at 1-6C has good accuracy because the cluster is centered, but the bullet holes are all over the paper, which indicates a lack of precision. The target in 1-6D lacks both accuracy and precision.

Target shooting is a good analogy for measurement processes and how to solve problems with them. Good shooting instructors know that it's better to work on precision first, that is, getting the cluster smaller (called "grouping" in shooting). That's analogous to reducing the random or inherent variation in a measurement process. Some of the clustering is due to the mechanics of the gun, but most of it happens to be due to the shooter. Once the shooter is consistently shooting tight clusters, then it's time to worry about moving the impact point (i.e., moving the average). How is this done? Not by adjusting the shooter, but adjusting the gun. The gun that shot the target at Figure 1.6B can be brought into good working order by adjusting the movable sights about two clicks to the right and two clicks down.

A mistake often made by novice shooters, as well as poor engineers or technicians, is to adjust the sights with too few shots on the paper. Enough shots (data points) must be collected to truly see the clustering before any meaningful change can be made. I have observed shooters firing two shots and then adjusting the sights; fire two more shots and adjust; fire two more shots and adjust; fire. . . . They wonder why the gun never seems to come into regulation. They blame the gun, the

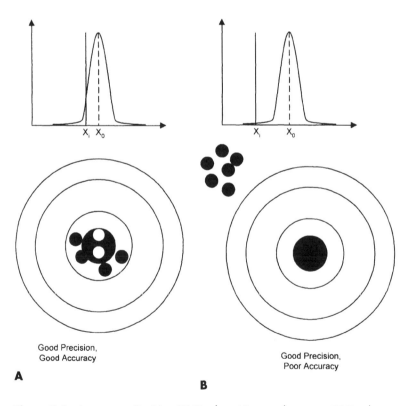

Figure 1.6 Accuracy vs. Precision: (A) Good precision, good accuracy; (B) Good precision, poor accuracy; (C) Poor precision, good accuracy; (D) Poor precision, poor accuracy.

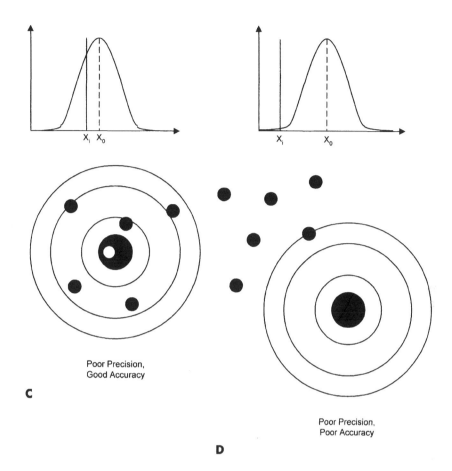

ammunition, and the lighting on the range—but never their methods. Bad science! And no art.

The *standard deviation* of the measurement is a good indication of its precision, which also means the inherent error in the measurement.

There are several tactics that help reduce the effects of error on practical measurements:

1. Make the measurement a large number of times, and then average the results.
2. Make the measurement several times using different instruments, if feasible.
3. When using instruments such as rulers or analog meters, try making the successive measurements on different parts of the scale. For example, on rulers and analog meter dials, the distance between tick marks is not really constant because of manufacturing error. The same is also true of electrical meter scales. Measure lengths using different points on the scale as the zero reference point (e.g., on a meter stick use 2, 12, 20, and 30 cm as the zero point), and then average the results. By taking the measurements from different sections of the scale, both individual errors and biases that accumulate will be averaged to a lower overall error.

Resolution

This term refers to the degree to which the measurand can be broken into identifiable adjacent parts. An example can be seen on the standard television test pattern broadcast by some stations in the early morning hours between "broadcast days." Various features on the test pattern will include parallel vertical or horizontal lines of different densities. One patch may be 100 lines per inch, another 200 lines per inch, and so forth up the scale. The resolution of the video system is the maximum density *at which it is still possible to see adjacent lines with space between them.* For any system, there is a limit above which the lines are blurred into a single entity.

The practical effects of resolution problems in a measurement situation are shown in Figure 1.7. In this case, some form of sensor is being used to detect two objects that are close together (X). The sensor is moved along a path parallel to the line of centers between target A and target B.

The dotted lines in Figure 1.7 show the approximate "field of view" of the sensor. If the sensitivity is set correctly, then we will have a response region that is smaller than the overall curve, so resolution is improved somewhat. But note that the field of view of Figure 1.7A is much broader than that of Figure 1.7B. The sensor in Figure 1.7A, therefore, has considerably worse resolution than that in Figure 1.7B. When these sensors look at the two targets, both targets appear in the field of view of Figure 1.7A at the same time, so the apparent target appears as one oblong blob. But in the higher resolution sensor, the two apparent targets are still oblong, but they are smaller and there is space visible between them.

In a digital electronic measuring instrument, the resolution is set by the number of "*bits*" (binary digits) used in the data word. Digital instruments use the binary (base-2) numbers system in which the only two permissible digits are "0" and "1." The binary "word" is a binary number representing a quantity. For example, binary "0001" represents decimal 1, while binary "1001" represents decimal 5. An 8-bit data word, the standard for many small embedded computers, can take on values from 00000000_2 to 11111111_2, so it can break the range into 2^8 (256) distinct

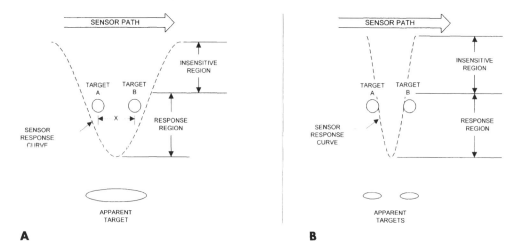

A   **B**

Figure 1.7   Resolution problem: (A) Both targets inside response curve results in poor resolution; (B) Narrower response curve breaks out the targets.

values, or $2^8 - 1$ (255) different segments. The resolution of that system depends on the value of the measured parameter change that must occur in order to change the least significant bit in the data word. For example, if 8 bits are used to represent a voltage range of 0 to 10 volts, then the resolution is $(10 - 0)/255$, or 0.039 V (e.g., 39 mV) per bit. This resolution is often specified as the "1-LSB" resolution of the instrument.

MEASUREMENT ERRORS

No measurement is perfect, and measurement apparatus are never ideal, so there will always be some *error* in all forms of measurement. An error is a deviation between the actual value of a measurand and the indicated value produced by the sensor or instrument used to measure the value. *Error is inherent and is not the fault of the person making the measurement.* Error is not the same as a mistake! Understanding error can greatly improve our effectiveness in making measurements.

Error can be expressed in either "absolute" terms or using a relative scale. An absolute error would be expressed in terms of "X-Volts ± Y-mV," or some other such unit, while a relative error expression would be "X ± 1% Volts." In an electrical circuit, a voltage might be stated as "4.5 volts ± 1%" (or "4.5 Volts ± 0.010 volts"). Which expression to use may be a matter of custom, convention, personal choice, or best utility, depending on the situation.

There are four general categories of error: *theoretical error, static error, dynamic error,* and *instrument insertion error.*

Theoretical Error

All measurements are based on some measurement theory that predicts how a value will behave when a certain measurement procedure is applied. The measurement theory is usually based on some theoretical model of the phenomenon being measured, or an intellectual construct that tells us something of how that phenomenon works. It is often the case that the theoretical model is valid only over a specified range of the phenomenon. For example, nonlinear phenomena that have a quadratic, cubic, or exponential function can be treated as a straight line linear function over small, carefully selected sections of the range. This would be a first-order approximation. Electronic sensor outputs often fall into this class.

Alternatively, the actual phenomenon may be terribly complex, or even chaotic, under the right conditions, so the model is therefore simplified for many practical measurements. An equation that is used as the basis for a measurement theory may be only a first-order approximation of the actual situation. For example, consider an example from the medical world: the *mean arterial pressure* (MAP) that is often measured in clinical medicine and medical sciences research situations. The MAP approximation equation used by clinicians is:

$$\overline{P} = Diastolic + \frac{Systolic - Diastolic}{3} \qquad [1.1]$$

This equation is really only an approximation (and holds true mostly for well people, not some sick people on whom it is applied) of the equation that expresses the mathematical integral of the blood pressure over a cardiac cycle: that is, the time average of the arterial pressure. The actual expression is written in the notation of calculus, which is beyond the math abilities of many of the people who use the clinical version above:

$$\bar{P} = \frac{1}{T} \int_{t1}^{t} 2P(t)dt \qquad [1.2]$$

The approximation works well, but is subject to greater error than the actual function due to the theoretical simplification of the first equation. What's funny about the first-order approximation of MAP blood pressure is that it often fails most on a very specific group of patients: *sick people!*

Static Errors

Static errors include a number of different subclasses that are all related in that they are always present even in unchanging systems (thus, they are not dynamic errors). These errors are not functions of the time or frequency variation.

Reading Static Errors

These errors result from misreading the display output of the sensor system. An analog meter uses a pointer to indicate the measured value. If the pointer is read at an angle other than straight on, then a *parallax reading error* occurs. Another reading error is the *interpolation error,* an error made in judging or estimating the correct value between two calibrated marks on the meter scale (Figure 1.8). Still another reading error occurs if the pointer on a meter scale is too broad and covers several marks at once.

A related error seen in digital readouts is the *last digit bobble error,* or the ± 1-count error. On digital displays, it is often the case that the least significant digit on the display will flip back and forth between two values. For example, a digital voltmeter might read "12.24" and "12.25" alternately, depending on when you looked at it, despite the fact that absolutely no change occurred in the voltage being measured. This phenomenon occurs when the actual voltage is midway between the two indicated voltages. Error, noise, and uncertainty in the system will make a voltage close to 12.245 volts bobble back and forth between the two permissible output

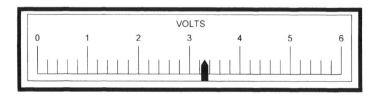

Figure 1.8   Analog meter indicating reading resolution problem.

states (12.24 and 12.25) on the meter. An example where "bobble" is of significant concern is the case where some action is taken when a value changes above or below a certain quantity—and the digital display bobbles above and below the critical threshold.

Environmental Static Errors

All sensors and instruments operate in an environment, which sometimes affects the output states. Factors such as temperature (perhaps the most common error-producing agent), electromagnetic fields, and radiation must be considered in electronic measurement systems.

Characteristic Static Errors

These static errors are still left after reading errors and environmental errors are accounted for. When the environment is well within the allowable limits and is unchanging, and when there is no reading error, there will be a residual error remaining that is a function of the measurement instrument or process itself. Errors found under this category include zero offset error, gain error, processing error, linearity error, hysteresis error, repeatability error, resolution error, and so forth. Also included in the characteristic error is any design or manufacturing deficiencies that lead to error. Not all of the "ticks" on the ruler are truly 1.0000 mm apart at all points along the ruler. While it is hoped that the errors are random, so that the overall error is small, there is always the possibility of a distinct bias or error trend in any measurement device.

For digital systems one must add to the resolution error a *quantization error* that emerges from the fact that the output data can only take on certain discrete values. For example, an 8-bit analog-to-digital converter allows 256 different states, so a 0 to 10 volt range is broken into 256 discrete values in 39.06 mV steps. A potential that is between two of these steps is assigned to one or the other according to the rounding protocol used in the measurement process. An example is the weight sensor that outputs 8.540 volts, on a 10-volt scale, to represent a certain weight. The actual 8-bit digitized value may represent 8.502, 8.541, or 8.580 volts because of the ± 0.039 volt quantization error.

Dynamic Errors

Dynamic errors arise when the measurand is changing or in motion during the measurement process. Examples of dynamic errors include the inertia of mechanical indicating devices (such as analog meters) when measuring rapidly changing parameters. There are a number of limitations in electronic instrumentation that fall into this category, especially cases where a frequency, phase, or slew rate limitation is present.

Instrument Insertion Error

The fundamental rule of making measurements is that *the measurement process should not significantly alter the phenomenon being measured.* Otherwise, the measurand is actually the altered situation, not the original situation that is of true interest. Examples of this error are found in many places. For example, a voltmeter with a low impedance of its own could alter resistance ratios in an electrical circuit and produce a false reading. This problem is especially seen when using cheap analog volt-ohm-milliammeters that have a low sensitivity, hence a low impedance, to measure a voltage in a circuit. The meter resistance R_m is effectively shunted across the circuit resistance across which the voltage appears.

Instrument insertion errors can usually be minimized by good instrument design and good practices. No measurement device has zero effect on the system being measured, but one can reduce the error to a very small value by appropriate selection of methods and devices.

DEALING WITH MEASUREMENT ERRORS

Measurement error can be minimized through several methods, some of which are lumped together under the rubric "procedure" and others under the legend "statistics."

Under "procedure" one can find methods that will reduce, or even minimize, error contributions to the final result. For example, in an electrical circuit, use a voltmeter that has an extremely high input impedance compared with circuit resistances. The idea is to use an instrument that least disturbs the thing being measured.

A significant source of measurement error in some electronic circuits is ground loop voltage drops and ground plane noise. This voltage may add or subtract from the reading of output voltage V_o, depending on its phase and polarity.

A way to reduce total error is to use several different instruments to measure the same parameter. In Figure 1.9 we see an example where the current flow in a circuit is being measured by three different ammeters: M1, M2, and M3. Each of these instruments will produce a result that contains a small error term decorrelated from the error of the others and not biased (unless, by selecting three identical model meters, we inherit some characteristic error of that particular type of instrument). We can estimate the correct value of the current flow rate by taking the average of the three:

$$\bar{I}_o = \frac{I_1 + I_2 + I_3}{3} \qquad [1.3]$$

One must be careful to either randomize the system in cases where the sensor or instruments used tend to have large error terms biased in one direction, or calibrate the average error so that it may be subtracted out of the final result.

Error Contributions Analysis

An *error analysis* can be performed in order to identify and quantify all contributing sources of error in the system. A determination is then made regarding the randomness of those errors, and a *worst case analysis* is made. Under the worst case, one assumes that all of the component errors are biased in a single direction and

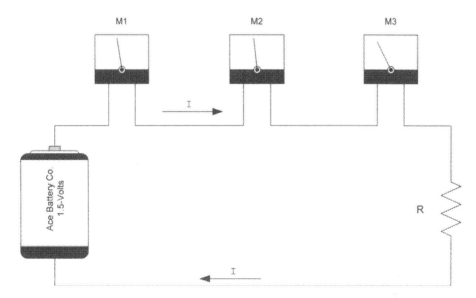

Figure 1.9 Summing measurements from multiple instruments improves probable accuracy.

are maximized. We then attempt to determine the consequences (for our purpose of making the measurement) if these errors line up in that manner, even if such an alignment is improbable. The worst case analysis should be done on both the positive and negative side of the nominal value. An *error budget* is then created to allocate an allowable error to each individual component of the measurement system in order to ensure that the overall error is not too high for the intended use of the system.

If errors are independent of each other ("decorrelated"), random rather than biased in one direction, and are of the same order of magnitude, then one can find the *root of the sum of the squares* (RSS) value of the errors. The RSS value can be used as a composite error term in planning a measurement system. The RSS error is:

$$E_{RSS} = \sqrt{\Sigma \varepsilon_i^2} \qquad [1.4]$$

Where: E_{RSS} is the composite error term, and ε_i represents the individual values of error. The RSS error E_{RSS} term is a reasonable estimate or approximation of the combined effects of the individual error components.

A collection of repetitive measurements of a phenomenon can be considered a *sampled population* and treated as such. If we take N measurements (M1 through M_n) of the same parameter and then average them, we get:

$$\overline{M} = \frac{M1 + M2 + M3 + \ldots + M_n}{N} \qquad [1.5]$$

The average value obtained in the equation above is the *mean arithmetic average*. This value is usually reported as the correct value for the measurement, but that when taken alone does not address the issue of error. For this purpose we add a correction factor by quoting the *standard error of the mean*, or

$$\sigma_{\overline{m}} = \frac{\sigma_m}{\sqrt{N}} \qquad [1.6]$$

Which is reported in the result as:

$$M = \overline{M} \pm \sigma_{\overline{m}} \qquad [1.7]$$

Any measurement contains error. This procedure allows us to estimate that error, and thereby understand the limitations of that particular measurement.

OPERATIONAL DEFINITIONS IN MEASUREMENT

Some measurement procedures suggest themselves immediately from the nature of the phenomenon being measured. In other cases, however, there is a degree of ambiguity in the process, and it must be overcome. Sometimes the ambiguity results from the fact that there are many different ways to define the phenomenon, or perhaps no single way is well established. In cases such as these, one might wish to resort to an *operational definition*, a procedure that will produce consistent results from measurement to measurement, or when measurements are taken by different people.

An operational definition, therefore, is a defined, standardized procedure that must be followed; it also specifies as many factors as are needed to control the measurement so that changes can be properly attributed only to the unknown variable. The need for operational definitions (as opposed to absolute) definitions arises from the fact that things are only rarely so neat, clean, and crisp as to suggest their own natural definition. By its very nature, the operational definition does not ask "true" or "false" questions, but rather it asks, "what happens under given sets of assumptions or conditions." What an operational definition can do for you, however, is to standardize a measurement in a clear and precise way so that it remains consistent across numerous trials. Operational definitions are used extensively in science and technology. When widely accepted, or promulgated by a recognized authority, they are called *standards*.

When an operational definition becomes widely accepted, and is used throughout an industry, it may become part of a formal *standard* or test procedure. You may, for example, see a procedure listed as "performed in accordance with NIST XXXX.XXX" or "ANSI Standard XXX," or "IEEE Standard XYZ." These notations mean that whoever made the measurement followed one or another of a published standard.

CHAPTER TWO

Small Components Used in Radio Frequency Test and Measurement

A number of small components are needed to do radio frequency (RF) tests and measurements. In this chapter we will look at some of the more common devices that should be in the kit of any workspace devoted to this class of testing.

ATTENUATORS

An *attenuator* is a device that reduces the amplitude of an applied signal. Figure 2.1A shows an attenuator in block form. A real attenuator may or may not have a ground connection, depending on whether it is *balanced* (no ground) or *single-ended* (grounded). Examples of both balanced and unbalanced types are shown below. Most commercially available attenuators are shielded, and the shield will be grounded even if the circuit inside is balanced.

Attenuation Definition

In an attenuator $P_{OUT} < P_{IN}$ by definition. The amount of attenuation can be expressed in either linear terms or in decibel notation, with the latter being more common. The attenuation factor in dB is:

$$Attenuation\ (dB) = 10\ Log\left[\frac{P_{OUT}}{P_{IN}}\right] \qquad [2.1]$$

Where:

      Attenuation (*dB*) is the reduction of input signal

in decibels

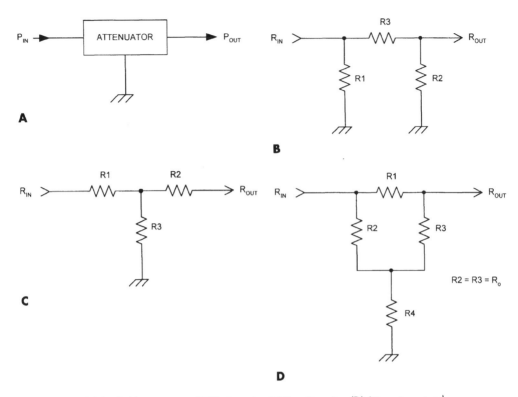

Figure 2.1   (A) Symbol for attenuator; (B) Pi-attenuator; (C) Tee-attenuator; (D) Attenuator network.

P_{OUT} is the output signal power level
P_{IN} is the input signal power level
Log refers to the base-10 logarithms

Note: P_{IN} and P_{OUT} are expressed in the same units (watts, milliwatts, microwatts).

Equation 2.1 gives the attenuation in terms of signal power level. To find the attenuation expressed in terms of input and output voltages or currents, replace the factor "10" with "20" in Equation 2.1. If you have the attenuation figure in dB, then the assumption is that it is relative to power levels. To find the voltage attenuation knowing the power attenuation, simply multiply power dB by two. Thus, a −3 dB attenuator will also have a −6 dB attenuation of voltage.

Impedances

The input and output impedances of attenuators of the attenuator are critical. Although circuits exist in which $Z_{IN} \neq Z_{OUT}$, the normal case is that $Z_{IN} = Z_{OUT}$. In most RF systems $Z_{IN} = Z_{OUT} = 50\text{-}\Omega$, but in television and video work the usual rule is $Z_{IN} = Z_{OUT} = 75\text{-}\Omega$ for unbalanced applications and 300-Ω for balanced cases. When using an attenuator to perform tests and measurements it is necessary

TABLE 2.1   Resistor values for various pi-attenuator networks.

Attenuation (dB)	R1 (Ohms)	R2 (Ohms)	R3 (Ohms)
1	870	870	5.8
2	436	436	11.6
3	292	292	17.6
6	150.5	150.5	37.3
10	96.2	96.2	71.2
12	83.5	83.5	93.2
20	61	61	247.5
30	53.2	53.2	789.7
40	51	51	2500
50	50.3	50.3	7905.6
60	50.1	50.1	25,000.00

to use a unit of the correct input/output impedance. Otherwise, reflections will occur because of the impedance mismatch, and the marked attenuation factor is incorrect. In the case of an impedance mismatch there will be a loss based on the VSWR, which in turn is the ratio of the impedances. For example, if you connect a 50-Ω attenuator into a 75-Ω circuit, the VSWR will be 75 Ω/50 Ω = 1.5:1.

Attenuator Examples

Figures 2.1B through 2.1D are examples of unbalanced fixed attenuators. The circuit in Figure 2.1B is a pi-network attenuator (named after the similarity to the Greek letter π). This circuit is probably the most common. Figure 2.1C is a Tee-network attenuator, while Figure 2.1D is a somewhat more complex attenuator circuit.

Table 2.1 shows the common values for the resistors in the Pi-attenuator of Figure 2.1B. In some cases, you might wish to build an attenuator, but normally that is not a wise use of time, especially given the fact that the resistor values are so specific. Commercial attenuators are available at very low cost from companies such as *Mini-Circuit Laboratories.*

Figure 2.2 shows the equivalent fixed attenuator circuits for balanced use. The version in Figure 2.2A is the balanced pi-attenuator, while that in Figure 2.2B is an H-pad attenuator. These circuits are used with balanced applications such as some television receivers. In the case of TV receivers, the balanced antenna scheme is normally 300-Ω rather than 75-Ω.

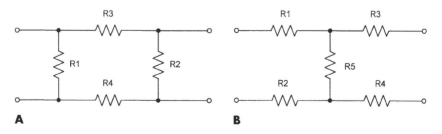

Figure 2.2   (A) Balanced pi-attenuator; (B) H-pad attenuator.

Switchable Attenuators

It is sometimes necessary to be able to switch attenuators in and out of a circuit. Figure 2.3 shows a typical circuit in which a DPDT toggle switch is used to select either the attenuator or the pass-through path. Note that there is a shield around the circuit. It is very important to include the shield in order to prevent leakage around the attenuator from creating levels that are not accounted for by the measuring process. Note that not all switches are well shielded, so be careful when selecting a switch type for S1.

Laboratory Attenuators

It is unlikely that building an attenuator is a worthwhile use of your time and resources. Figure 2.4 shows several types of commercial attenuators used in the laboratory. A fixed attenuator is shown in Figure 2.4A. This device is sometimes called a *barrel attenuator* or *in-line attenuator*. It has a male BNC on one end and a female BNC on the other (other connector combinations are also available, but BNC is the most common).

A *step attenuator* is shown in Figure 2.4B. It consists of a series of circuits such as Figure 2.3, each in its own shielded compartment within the shielded enclosure for the entire device. In this version female BNC connectors are on either end, while the toggle switches are on the top. To select an amount of attenuation, the required switches are turned on or off as needed. For example, to make a 3-dB attenuator, turn on the 1-dB and 2-dB switches and leave all others turned off.

An electrically switchable step attenuator is shown in Figure 2.4C. This device has feed-through or EMI filtering capacitors mounted on the shielded enclosure. Applying a DC voltage to the connector on the capacitor turns on the associated attenuator. In some cases, electronic switching is used inside the box, but in others electromechanical relays are used.

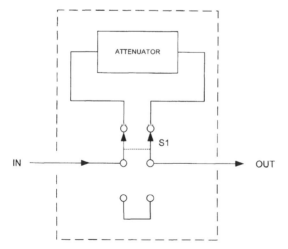

Figure 2.3
Switchable attenuator.

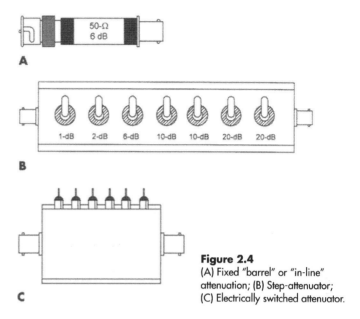

Figure 2.4
(A) Fixed "barrel" or "in-line" attenuation; (B) Step-attenuator; (C) Electrically switched attenuator.

DUMMY LOADS

A dummy load is a resistive load with a specified resistance value. It is used to terminate ports that are unused, as well as in certain other cases. Figure 2.5A shows the basic circuit for a dummy load. It consists of a noninductive resistor inside a shielded enclosure, with a coaxial connector (BNC shown) to provide access.

It is very important that the resistor (R) have a value equal to the desired system impedance (e.g., 50 Ω) and be noninductive. The latter requirement means that wire-wound resistors are not usable. Carbon composition or metal film resistors are normally used for this purpose. Standard values of resistor include 51 ohms, and for many purposes this is sufficient. It is possible to buy 50-Ω precision resistors. It is also possible to use multiple resistors (e.g., two 100 Ω or three 150 Ω in parallel).

The power rating of the dummy load should be sufficient to prevent resistance changes due to self-heating. In many cases, a half-watt dummy load is usable, but if significant amounts of RF power are anticipated, a higher rating is required. When transmitters are tested it is common practice to use a dummy load to absorb the RF power generated. These dummy loads might be able to dissipate power levels from 1-watt to many kilowatts depending on the nature of the transmitter being tested. Any of these dummy loads can also be used for other forms of testing at lower power levels. Figure 2.5B shows a dummy load intended for testing small VHF-FM marine transmitters. It can be used for laboratory testing as well, and is small enough to be practical.

Figure 2.5C shows a shop-built switchable dummy load. It permits the selection of a number of different resistances, as well as providing a pass-through position that eliminates the dummy load altogether.

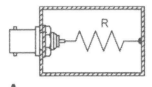

A

B

C

Figure 2.5 (A) Dummy load circuit; (B) Commercial low-power dummy load; (C) Switched load.

BALUN AND IMPEDANCE MATCHING TRANSFORMERS

There are at least two reasons why you might want to use a transformer in an RF test or measurement set-up. One is to match impedances between two different devices. Another is to convert between balanced and unbalanced circuits. In the latter case, a BALUN (BALanced-UNbalanced) transformer is used.

Figure 2.6 shows several different transformer styles. The transformer shown in Figure 2.6A is a BALUN with a 1:1 impedance ratio. It is used to provide translation between balanced and unbalanced circuits, but without impedance transformation. The version in Figure 2.6B is also a BALUN transformer, but provides a 4:1 impedance transformation.

The transformers in Figures 2.6C and 2.6D are both unbalanced on both ends (some call them "UN-UNs"). The transformer of Figure 2.6C provides a 9:1 impedance transformation, while Figure 2.6D provides 16:1 transformation. These circuits are predicated on using an equal number of turns on each winding. Other transformation ratios can be created using different turns ratios, or by tapping a single winding.

Figure 2.6E shows how these transformers are constructed. The wire is wound on a ferrite or powdered iron toroidal-shaped core using either the *bifilar* (two windings, as shown in Figure 2.6E), *trifilar* (three windings), or *multifilar* (any number of windings) approach.

COAXIAL SWITCHES

If it is necessary to switch between devices or loads, then it might be prudent to use a *coaxial switch box*, such as the example shown in Figure 2.7. This particular device was designed for amateur radio operators, so the labeling reflects that application. However, it is also useful for other RF switching purposes.

RF POWER COMBINERS AND SPLITTERS

The principal difference between power combiners and power splitters is in the application. Otherwise, they are the same circuits. A *combiner* is used whenever it is necessary to mix two or more signals sources linearly into a common port. The combiner is not a mixer because it is linear, and thus does not produce additional frequency products. The *splitter* performs exactly the opposite function. It will direct RF power from a single source to two more loads.

Resistive Combiner/Splitter

Perhaps the simplest circuit is the resistive network in Figure 2.8. This circuit uses three resistors in a Y-network to provide three ports (it can also be extended to higher numbers of ports). The value of each resistor is $R = R_0/N$, where R_0 is the

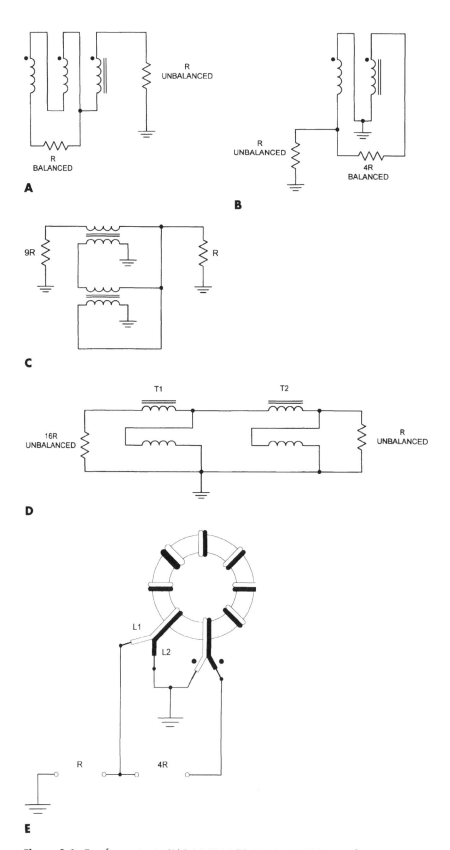

Figure 2.6 Transformer circuits: (A) 1:1 BALUN; (B) 4:1 BALUN; (C) 9:1 transformer; (D) 16:1 transformer; (E) bifilar winding.

Figure 2.7 Coaxial switch.

system impedance and N is the number of ports. For example, if the system impedance is 50 ohms, then R = 16.67 ohms, and for 75 ohm systems R = 25 ohms.

The resistors used in this circuit must be *noninductive*. This limits selection to carbon composition or metal film resistors. If higher power than 2-watts is needed, then each arm of the Y-network must be made from multiple resistors in series or parallel. The values 16.67 ohms and 25 ohms are not standard values, except perhaps in certain lines of 1% or less tolerance precision resistors. They can, however, be approximated using standard values. For example, only a small error is created when 15 ohm resistors are used in place of 16.67 ohms, and 27 ohms is used in place of 25 ohms. Because resistors come with variation in actual value, the amount of which is indicated in its tolerance rating (5%, 10%, 20%), we can often select from a collection of standard values to closely approximate the actual value needed.

It is also possible to approximate the values by using series or parallel combinations of standard value resistors. For example, a pair of 51-ohm standard-value resistors in parallel will make a good match for 25 ohms. Similarly, three 51-ohm resistors in parallel will closely approximate 16.67 ohms.

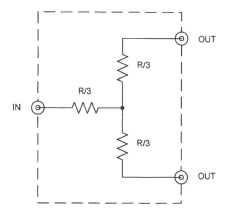

Figure 2.8
Resistive combiner/splitter.

The advantages of the resistive combiner/splitter are its broadband operation. The bandwidth can extend into the UHF region with discrete resistors, and into the gigahertz region if implemented with surface-mount resistors and appropriate printed circuit technology. The stray inductance and capacitance set the upper frequency limit in either case.

The disadvantages of this form of combiner include a relatively high insertion loss, 6 dB, of which 3 dB is due to the resistors, and 3 dB is due to the fact that the input power is split two ways. Isolation between output ports is about 6 dB. If those can be overcome, or are not important in a given application, then this form of splitter/combiner is ideally suited.

Transformer Combiner/Splitter

Figure 2.9 shows a somewhat better form of combiner/splitter circuit. This circuit can be used from 500 kHz to over 1,000 MHz if the proper transformers and capacitor are provided. This discussion will concentrate on the high frequency shortwave bands, as those are the easiest forms of combiner/splitter for most readers to actually build.

The power-splitting function is performed by coil L2. This coil is center-tapped, with the input signal applied to the tap and the outputs taken from the ends. This transformer can be wound on either T-50-2 or T-50-6 toroidal cores for the HF bands, or a T-50-15 core for the AM BCB and medium wave bands. Use 18 turns of #26 AWG wire for the HF bands, and 22 turns for MW bands.

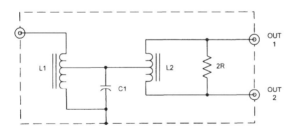

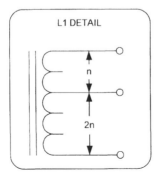

Figure 2.9
Transformer combiner/splitter.

The resistor across the ends of L2 should be twice the system impedance. That means 100 ohms for 50-ohm systems, and 150 ohms for 75-ohm systems (both are standard values).

Some impedance transformation is needed if the system impedance is to be maintained, so L1 must be provided. This transformer is tapped, but not at the center. The inset detail in Figure 2.9 shows the relationship of the tap to the winding: it is located at the one-third point on the winding. If the bottom of the coil is grounded, then the tap is at the two-thirds point (2N turns), and the input is at the top (N + 2N turns). In other words, the tap is at two-thirds the overall length of the winding.

The capacitor usually has a value of 10-pF, although people with either a sweep generator—or a CW RF signal generator and a lot more patience than I have—can optimize performance by replacing it with a 15-pF trimmer capacitor. Adjust the trimmer for flattest response across the entire band.

It is important to use toroid core inductors for the combiner. The most useful core types are listed above, although for other applications other cores could also be used. Figure 2.10 shows one way in which the cores can be wound. This is the linear winding approach, that is, it uses a single coil of wire. The turns are wound until the point where the tap occurs. At that point one of two approaches is taken. First, you could end the first half of the winding and cut the wire. Adjacent to the tap, start the second half of the winding. Scrape the insulation off the ends at the tap, and then twist the two ends together to form the tap. Alternatively, you can loop the wire (see detail inset to Figure 2.10) and then continue the winding. The loop then becomes the tap. Scrape the insulation off the wire, and solder it. Although the tap here is a center tap (which means L2), it also serves for L1 if you offset the tap a bit to the left or right.

An alternate method for L2 is shown in Figure 2.11. This is superior to the other form for L2, but it is a little more difficult. Either wind the two wires together side-by-side (Figure 2.11A), or twist them together before winding (Figure 2.11B). Make a loop at the center-tap, and scrape it for soldering.

Modified VSWR Bridge Combiner/Splitter

Figures 2.12 and 2.13 show 6 dB combiner/splitter circuits based on the popular bridge used to measure voltage standing wave ratio (VSWR). Each of these circuits uses a bridge made of three resistors and one winding of a transformer. In both cases, the transformers have a 1:1 turns ratio. Also, in both cases $R1 = R2 = R3 = R_o$. In 50-ohm systems, therefore, the value of the resistors is 50 ohms, and in 75-ohm systems it is 75 ohms. In Figure 2.12 the transformer is not tapped. It is a straight 1:1 turns ratio toroid transformer. The circuit in Figure 2.13, however, uses a center tap on the primary of T1. Note that there is a difference in the location of the summation output between the two circuits. These circuits have been popular for combining two-signal generator (e.g., a sweep generator) and a marker generator.

A variation on the theme is shown in Figure 2.14. This circuit is sometimes also used as a directional coupler. RF power applied to the input port appears at OUT-2 with only an insertion-loss attenuation. A sample of the input signal appears at OUT-1. Alternatively, if RF power is applied to OUT-2, it will appear at the input, but does not appear at the OUT-1 port due to cancellation.

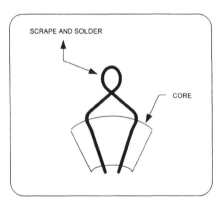

Figure 2.10
Toroid winding.

For the case where the device has a −3.3 dB output at OUT-2 and a −10 dB output at OUT-1, and a 50-ohm system impedance (R_0), the value of R1 = 108 ohms, and R2 = 23 ohms. The equations for this device are:

$$R_o = \sqrt{R1\ R2} \qquad [2.2]$$

$$C.F. = 20\ Log\left(\frac{R_o}{R1 + R_o}\right) \qquad [2.3]$$

$$L_i = -20\ Log\left(\frac{R_o}{R2 + R_o}\right) \qquad [2.4]$$

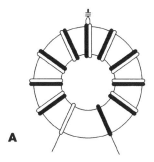

A

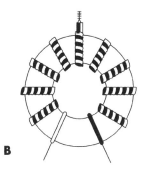

B

Figure 2.11
Two different bifilar winding styles.

Where:

R_0 is the system impedance (e.g., 50 Ω)
C.F. is the coupling factor
L_i is the insertion loss from IN to OUT-2
R1 and R2 are the resistances of R1 and R2

Taking the negative of Equation 2.2 gives the insertion loss from IN to OUT-1.

90° Splitter/Combiner

Figure 2.15 shows a 3 dB splitter/combiner made of lumped L and C elements, which produces a 0° output at OUT-1 and a 90° output at OUT-2. A closely coupled 1:1 transformer is used to supply two inductances, L1 and L2. This transformer is wound in the bifilar manner to ensure tight coupling. The values of

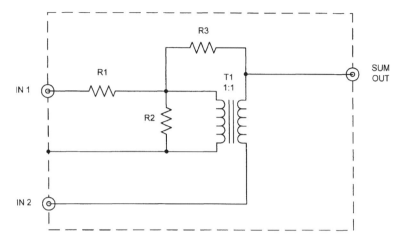

Figure 2.12   VSWR bridge combiner/splitter.

inductance and capacitance, assuming that L1 = L2 = L, and C1 = C2 = C, are given by:

$$L = \frac{R_o}{2.828\, f_{3dB}} \quad\quad [2.5]$$

$$C = \frac{1}{2.828\, f_{3dB}\, R_o} \quad\quad [2.6]$$

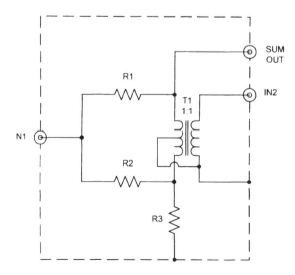

Figure 2.13
Alternate VSWR combiner/splitter.

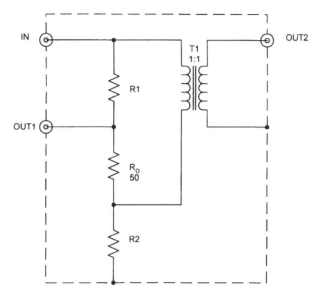

Figure 2.14
Unequal output combiner/splitter.

Where:

L is the inductance of L1 and L2
C is the capacitance of C1 and C2
R_0 is the system impedance (e.g., 50 ohms)
f_{3dB} is the 3 dB coupling frequency

The bandwidth of this circuit is approximately 20 percent for 1 dB amplitude balance.

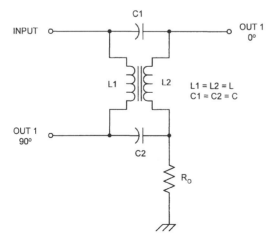

Figure 2.15
Quadrature combiner/splitter.

Transmission Line Splitter/Combiner

The Wilkinson power splitter/combiner is shown in Figure 2.16. This network can achieve 2-dB isolation between the two output ports over a bandwidth that is approximately ±20% of the design frequency. It consists of two transmission lines, TL1 and TL2, and a bridging resistor (R), which has a value of $R = 2R_o = (2)(50 \, \Omega) = 100 \, \Omega$.

Transmission lines TL1 and TL2 are each quarter wavelength and have a characteristic impedance equal to 1.414 times the system impedance. If the system impedance is 50 ohms, then the value of the characteristic impedance needed for the transmission lines $(1.414)(50 \, \Omega) = 70.7 \, \Omega$.

The Wilkinson network can be implemented using coaxial cable at VHF and below, although at higher frequencies printed circuit transmission line segments are required. If coaxial cable is used, then the physical length of TL1 and TL2 are shortened by the *velocity factor* (VF) of the cable used for TL1 and TL2. The values of VF will be 0.66 for polyethylene dielectric coax, 0.80 for polyfoam dielectric, and 0.70 for *Teflon*™ dielectric cable. The physical length is:

$$Length = \frac{75 \, VF}{F_{MHz}} \text{ meters} \qquad [2.7]$$

An N-way version of the same idea is shown in Figure 2.17. In this network a transmission line, TL1 − TL(n), and resistor are used in each branch. The resistor values are the value of R_o. In the case shown, the values of resistors are 50 ohms because it is designed for standard 50-ohm systems. The characteristic impedance of the transmission lines used in the network is:

$$Z_o = R_o \sqrt{N} \qquad [2.8]$$

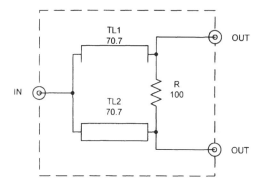

Figure 2.16
Wilkinson combiner/splitter.

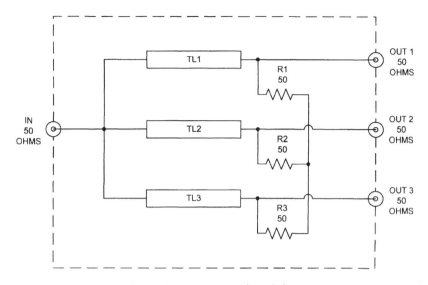

Figure 2.17 N-way combiner/splitter.

Where:

Z_0 is the characteristic impedance of the transmission lines

R_0 is the system standard impedance

N is the number of branches

In the case of a 50-ohm system with three branches, the characteristic imped-ance of the lines is $(50\ \Omega)(\sqrt{3}) = (50\ \Omega)(1.73) = 86.5\ \Omega$.

90° Transmission Line Splitter/Combiner

Figure 2.18 shows the network for producing 0°–90° outputs, with −3 dB loss, us-ing transmission line elements. The terminating resistor at one node of the bridge is the system impedance, R_0 (e.g., 50 ohms).

Each transmission line segment is quarter wavelength ($\lambda/4$), so they have physical lengths calculated from Equation 2.7 above. The characteristic impedance of TL1 and TL2 is the system impedance, R_0, while the impedance pf TL3 and TL4 are $0.707R_0$. In the case of 50-ohm systems, the impedance of TL3 and TL4 is 35 ohms.

HYBRID RING "RAT-RACE" NETWORK

The "Rat-Race" network of Figure 2.19 has a number of applications in communi-cations. It consists of five transmission line segments, TL1 through TL5. At VUF, UHF, and microwave frequencies this form is often implemented in printed circuit board transmission lines.

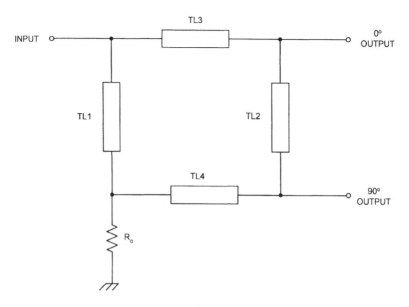

Figure 2.18   Transmission line quadrature combiner/splitter.

Four of these transmission line segments (TL1 to TL4) are quarter wavelength, while TL5 is half wavelength. The characteristic impedance of all lines is $1.414R_0$. Each quarter wavelength segment creates a 90° phase shift, while the half wavelength produces a 180° phase shift. When a quarter wavelength and half wavelength are combined, to form a three-quarter wavelength segment, the phase shift is 270°.

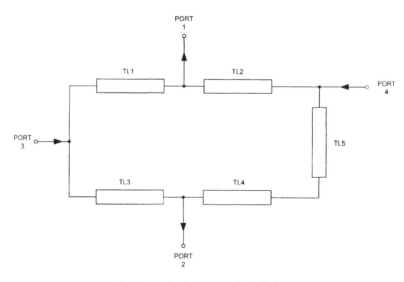

Figure 2.19   Rat-race combiner/splitter.

TABLE 2.2

Input	Use
Port 3	0° Splitter, −3 dB at Ports 1 and 2
Port 4	180° Splitter, −3 dB, −90° at Port 1, and −270° at Port 2

It is necessary to terminate all ports of the Rat-Race network in the system characteristic impedance, R_0, whether they are used or not. The bandwidth of this network is approximately 20%.

Different applications use different ports for input and output. Table 2.2 shows some of the relationships found in this network.

A coaxial cable version is shown in Figure 2.20. This network is implemented using coaxial cable sections and Tee connectors. In this case, there are three quarter-wavelength sections (90°) and one three-quarter wavelength (270°) section. Applications of this network include those where a high degree of isolation is required between ports.

THE RF HYBRID COUPLER

The hybrid coupler (Figure 2.21) is a device that will either split a signal source into two directions, or combine two signals sources into a common path. The circuit symbol shown in Figure 2.21 is essentially a signal path schematic. Consider the situation where an RF signal is applied to Port 1. This signal is divided equally, flowing to both Port 2 and Port 3.

Because the power is divided equally the hybrid is called a 3 dB divider, that is, the power level at each adjacent port is one-half (−3 dB) of the power applied to the input port.

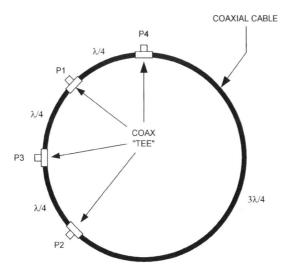

Figure 2.20
Coaxial rat-race combiner/splitter.

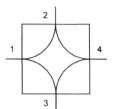

Figure 2.21
Symbol for hybrid.

If the ports are properly terminated in the system impedance, then all power is absorbed in the loads connected to the ports adjacent to the injection port. None travels to the opposite port. The termination of the opposite port is required, but it does not dissipate power because the power level is zero.

The one general rule to remember about hybrids is that *opposite ports cancel.* That is, power applied to one port in a properly terminated hybrid will not appear at the opposite port. In the case cited above, the power was applied to Port 1, so no power appeared at Port 4.

One of the incredibly useful features of the hybrid is that it accomplishes this task while allowing all devices connected to it to see the system impedance, R_o. For example, if the output impedance of the signal source connected to Port 1 is 50 ohms, the loads of Port 2 and Port 3 are 50 ohms, and the dummy load attached to Port 4 is 50 ohms, then all devices are either looking into, or driven by, the 50-ohm system impedance.

One source of reasonably priced hybrid devices is Mini-Circuits Laboratories (13 Neptune Avenue, Brooklyn, NY, 11235. Web site: http://www.minicircuits.com). They have a large selection of 0°, 90°, and 180° hybrid combiners and splitters.

Applications of Hybrids

The hybrid can be used for a variety of applications where either combining or splitting signals is required.

Combining Signal Sources

In Figure 2.22 there are two signal generators connected to opposite ports of a hybrid (Port 2 and Port 3). Power at Port 2 from Signal Generator No. 1 is therefore canceled at Port 3, and power from Signal Generator No. 2 (Port 3) is canceled at Port 2. Therefore, the signals from the two signal generators will not interfere with each other.

In both cases, the power splits two ways. For example, the power from Signal Generator No. 1 flows into Port 2 and splits two ways. Half of it (3 dB) flows through the path from Port 2 to Port 1, while the other half flows from Port 2 to Port 4. Similarly with the power from Signal Generator No. 2 applied to Port 3. It splits into two equal portions, with one flowing to Port 1 and the device under test, and half flowing to the dummy load.

Bidirectional Amplifiers

A number of different applications exists for bidirectional amplifiers, such as amplifiers that can handle signals from two opposing directions on a single line. The

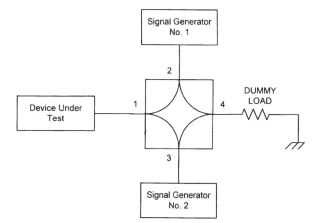

Figure 2.22
Combining two signal sources.

telecommunications industry, for example, uses such systems to send full duplex signals over the same lines.

Similarly, cable TV systems that use two-way (e.g., cable MODEM) require two-way amplifiers. Figure 2.23 shows how the hybrid coupler can be used to make such an amplifier. In some telecommunications textbooks the two directions are called East and West, so this amplifier is occasionally called an East-West (E-W) amplifier. At other times this circuit is called a *repeater.*

In the bidirectional E-W amplifier of Figure 2.23, amplifier A1 amplifies the signals traveling West-to-East, while A2 amplifies signals traveling East-to-West. In each case, the amplifiers are connected to hybrids HB1 and HB2 via opposite ports, so they will not interfere with each other. Otherwise, connecting two amplifiers input-to-output-to-input-to-output is a recipe for disaster—if only a large amount of destructive feedback.

Transmitter/Receiver Isolation

One of the problems that exists when using a transmitter and receiver together on the same antenna is isolating the receiver input from the transmitter input. Even a

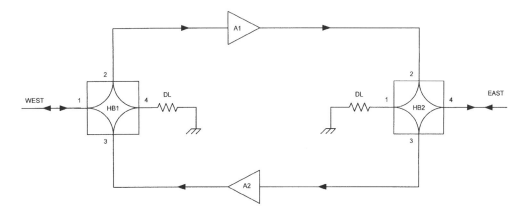

Figure 2.23 Bidirectional "repeater" amplifier.

weak transmitter will burn out the receiver input if its power were allowed to reach the receiver input circuits. One solution is to use one form of transmit/receive (T/R) relay. But that solution relies on an electromechanical device, which adds problems of its own (not the least of which is reliability). A solution to the T/R problem using a hybrid is shown in Figure 2.24. In this circuit the transmitter output and receiver input are connected to opposite ports of a hybrid device. Thus, the transmitter power does not reach the receiver input.

The antenna is connected to the adjacent port between the transmitter port and the receiver port. The signal from the antenna will flow over the Port-1-to-Port-2 path to reach the receiver input. Transmitter power, on the other hand, will enter at Port 3, and is split into two equal portions. Half the power flows to the antenna over the Port-3-to-Port-1 path, while half the power flows to a dummy load through the Port-3-to-Port-4 path.

There is a problem with this configuration. Because half the power is routed to a dummy load, there is a 3-dB reduction in the power available to the antenna. A solution is shown in Figure 2.25. In this configuration a second antenna is connected in place of the dummy load. Depending on the spacing (S), and the phasing, various directivity patterns can be created using two identical antennas.

If the hybrid produces no phase shift of its own, then the relative phase shift of the signals exciting the antennas is determined by the length of the transmission line between the hybrid and that antenna. A 0° phase shift is created when both transmission lines are the same length. Making one transmission line a half wavelength longer than the other results in a 180°-phase shift. These two relative phase relationships are the basis for two popular configurations of phased array antenna. Consult an antenna book for other options.

Phase-Shifted Hybrids

The hybrids discussed thus far split the power half to each adjacent port, but the signals at those ports are in-phase with each other. That is, there is a zero-degree

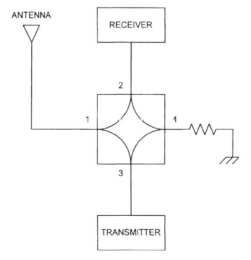

Figure 2.24
Use of hybrid as a T/R switch.

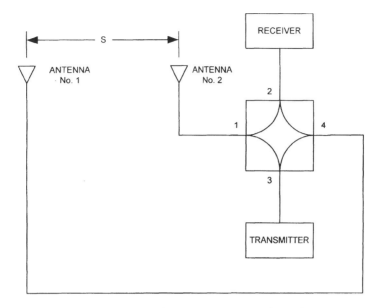

Figure 2.25 Combining two antennas in T/R switch.

phase shift over the paths from the input to the two output ports. There are, however, two forms of phase-shifted hybrids. The form shown in Figure 2.26A is a 0°–180° hybrid. The signal over the Port-1-to-Port-2 path is not phase shifted (0°), while that between Port 1 and Port 3 is phase shifted 180°. Most transformer-based hybrids are inherently 0° to 180° hybrids.

A 0° to 90° hybrid is shown in Figure 2.26B. This hybrid shows a 90°-phase shift over the Port 1/Port 2 path, and a 0° phase shift over the Port 1/Port 3 path. This type of hybrid is also called a *quadrature hybrid*.

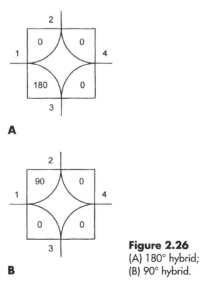

Figure 2.26
(A) 180° hybrid;
(B) 90° hybrid.

One application for the quadrature hybrid is the balanced amplifier shown in Figure 2.27. Two amplifiers, A1 and A2, are used to process the same input signal arriving via hybrid HB1. The signal splits in HB1, so it becomes inputs to both A1 and A2. If the input impedances of the amplifiers are not matched to the system impedance, then the signal will be reflected from the inputs back towards HB1. The reflected signal from A2 arrives back at the input in-phase (0°), but the one reflected from A1 has to pass through the 90°-phase shift arm twice, so it has a total phase shift of 180°. Thus, the reflections caused by mismatching the amplifier inputs are cancelled out.

The output signals of A1 and A2 are combined in hybrid HB2. The phase balance is restored by the fact that the output of A1 passes through the 0° leg of HB2, while the output of A2 passes through the 90° leg. Thus, both signals have undergone a 90°-phase shift, so they are now restored to the in-phase condition.

Use with Receive Antennas

The examples given above combine a receiver and transmitter on a single antenna or antenna system. It is also possible to use the hybrid for antenna arrays intended for receivers. Antenna spaced some distance (X) apart will have different patterns and gains depending on the value of X and the relative phase of the currents in the two antennas. One can, therefore, connect the antennas to Ports 2 and 3, and the receiver antenna input to Port 1. A terminating resistor would be used at Port 4. You can use either 0°, 90°, or 180° hybrids depending on the particular antenna system.

DIPLEXERS

The diplexer is a passive RF device that provides frequency selectivity at the output, while looking like a constant resistive impedance at its input terminal. Figure 2.28 shows a generalization of the diplexer. It consists of a *high-pass filter* and a *low-*

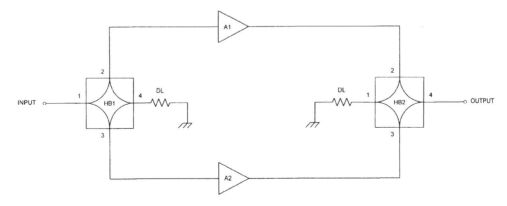

Figure 2.27 Balanced amplifier.

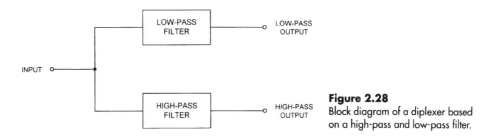

Figure 2.28
Block diagram of a diplexer based on a high-pass and low-pass filter.

pass filter that share a common input line. With appropriate design the diplexer will not exhibit any reactance reflected back to the input terminal (which eliminates the reflections and VSWR problem). Yet, at the same time it will separate the high and low frequency components into two separate signal channels.

Figure 2.29 shows a practical diplexer built with lumped constant inductor (L) and capacitor (C) components. The high-pass filter consists of C1, C2, C3, L1, and L2. The low-pass filter consists of C4, C5, L3, L4, and L5. Both filters share a common ground at the chassis.

The LC component values shown in Figure 2.29 are for a −3 dB cut-off frequency of 100 MHz for both the high-pass and low-pass filter sections. Note that the corner frequency of the filters is the same for both sections, so there is no portion of the spectrum that is not covered by the filter, except very close to the 100-MHz corner frequency. The circuit's component values can be scaled to other frequencies by using a simple formula:

$$New\ Value = \frac{Old\ Value \times 100}{F_{MHz}}$$ [2.9]

Where F_{MHz} is the desired frequency in megahertz (MHz).

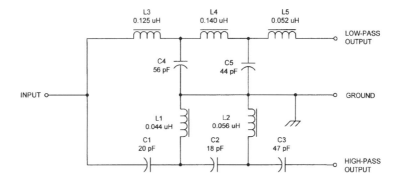

Figure 2.29   Low-pass/high-pass diplexer circuit centered on 100 MHz.

Using a Diplexer with a Mixer

One common use for the diplexer circuit is to smooth the impedance excursions seen by a mixer circuit, and to prevent reflected signals from the load back to the mixer from interfering with mixer performance. Reflections can seriously affect the intercept points of the mixer, which means that they can reduce the dynamic range and increase undesirable intermodulation distortion. The diplexer approach of Figure 2.30 is particularly useful where the mixer and the following circuitry are mismatched at frequencies other than the desired frequency.

Figure 2.30 shows the two cases. In each case, a mixer nonlinearly combines two frequencies, F1 and F2, to produce an output spectrum of $mF1 \pm nF2$, where m and n are integers representing the fundamental and harmonics of F1 and F2. In some cases, we are interested only in the difference frequency, so will want to use the low-pass output (LPO) of the diplexer (Figure 2.30A). The high-pass output (HPO) is terminated in a matched load so that signal transmitted through the high-pass filter is fully absorbed in the load.

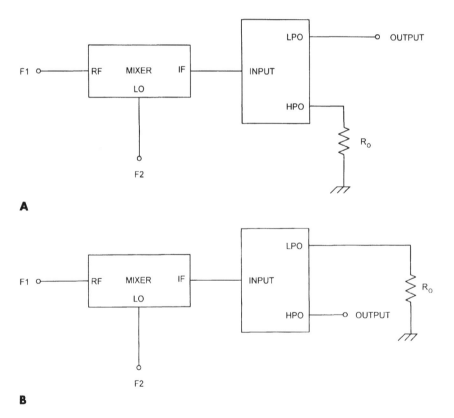

Figure 2.30 Use of a diplexer to improve mixer performance: (A) for difference IF; (B) for sum IF.

The exact opposite situation is shown in Figure 2.30B. Here we are interested in the sum frequency, so use the HPO port of the diplexer, and terminate the LPO port in a resistive load. In this case, the signal passed through the low-pass filter section will be absorbed by the load.

Bandpass Diplexers

Figures 2.31 and 2.32 show two different bandpass diplexer circuits commonly used at the outputs of mixers. These circuits use a bandpass-filter approach, rather than two separate filters. The version in Figure 2.31 is a π-network approach, while the version in Figure 2.32 is an L-network. In both cases,

$$Q = \frac{f_o}{BW_{3dB}} \qquad\qquad [2.10]$$

and,

$$\omega = 2\pi f_o \qquad\qquad [2.11]$$

Where:

f_o is the center frequency of the passband in hertz (Hz)
BW_{3dB} is the desired bandwidth in hertz (Hz)
Q is the relative bandwidth

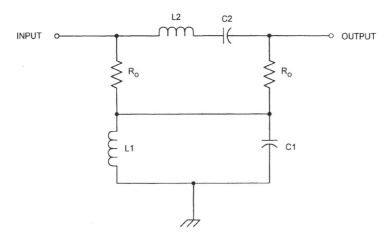

Figure 2.31 Pi-type bandpass diplexer.

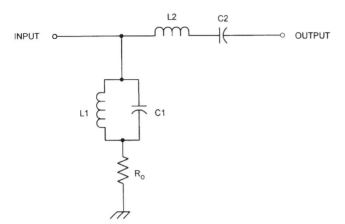

Figure 2.32
L-type bandpass diplexer.

For the circuit of Figure 2.31:

$$L2 = \frac{R_o Q}{\omega} \qquad [2.12]$$

$$L1 = \frac{R_o}{\omega Q} \qquad [2.13]$$

$$C2 = \frac{1}{R_o Q \omega} \qquad [2.14]$$

$$C1 = \frac{Q}{\omega R_o} \qquad [2.15]$$

For the circuit of Figure 2.32:

$$L2 = \frac{R_o Q}{\omega} \qquad [2.16]$$

$$L1 = \frac{R_o}{\omega Q} \qquad [2.17]$$

$$C1 = \frac{1}{L1\,\omega^2}$$ [2.18]

$$C2 = \frac{1}{L2\,\omega^2}$$ [2.19]

DIRECTIONAL COUPLERS

Directional couplers are devices that will pass a signal across one path, while passing a much smaller signal along another path. One of the most common uses of the directional coupler is to sample an RF power signal either for controlling transmitter output power level or for measurement. An example of the latter use is to connect a digital frequency counter to the low-level port, and the transmitter and antenna to the straight-through (high-power) ports.

A transmission line directional coupler is shown in Figure 2.33. The circuit symbol is shown in Figure 2.33A. Note that there are three outputs and one input. The IN-OUT path is low-loss and is the principal path between the signal source and the load. The coupled output is a sample of the forward path, while the isolated output showed very low signal. If the IN and OUT are reversed, then the roles of the coupled and isolated ports also reversed.

An implementation of this circuit using transmission line segments is shown in Figure 2.33B. Each transmission line segment (TL1 and TL2) has a characteristic impedance, Z_O, and is a quarter wavelength long. The path from Port 1 to Port 2 is the low-loss signal direction. If power flows in this direction, then Port 3 is the coupled port and Port 4 is isolated. If the power flow direction reverses (Port 2 to Port 1) then the respective roles of Port 3 and Port 4 reverse.

For a coupling ratio (Port 3/Port 4) ≤ -15 dB the value of coupling capacitance must be:

$$C_c < \frac{0.18}{\omega Z_o} \text{ farads}$$ [2.20]

The coupling ratio is:

$$C.R. = 20\,Log\,(\omega C Z_o)db$$ [2.21]

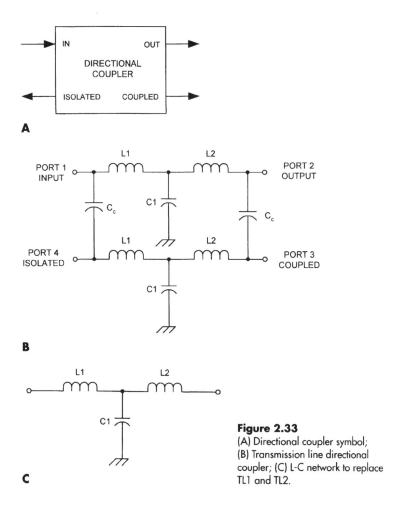

Figure 2.33
(A) Directional coupler symbol;
(B) Transmission line directional
coupler; (C) L-C network to replace
TL1 and TL2.

The circuit shown in Figure 2.33C is an LC lumped constant version of the transmission lines. This network can be used to replace TL1 and TL2 in Figure 2.33A. The values of the components are:

$$L1 = \frac{Z_o}{\omega_o} \qquad [2.22]$$

$$C1 = \frac{1}{\omega_o Z_o} \qquad [2.23]$$

Figure 2.34 shows a directional coupler used in a lot of RF power meters and VSWR meters. The transmission lines are implemented as printed circuit board tracks. It consists of a main transmission line (TL1) between Port 1 and Port 2 (the low-loss path), and a coupled line (TL2) to form the coupled and isolated ports. The

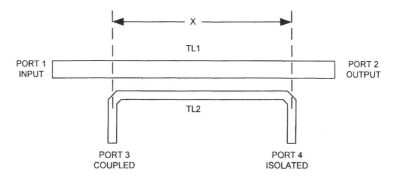

Figure 2.34   Printed circuit directional coupler.

coupling capacitance (in pF) is approximated by 9.399X when implemented on G-10 Epoxy fiberglass printed circuit board.

A reflectometer directional coupler is shown in Figure 2.35A. This type of directional coupler is at the heart of many commercial VSWR meters and RF power meters used in the HF through low-VHF regions of the spectrum. This circuit is conceptually similar to the previous transmission line, but it is designed around a toroid transmission line transformer. It consists of a transformer in which the low-loss path is a single-turn primary winding and a secondary wound of enameled wire.

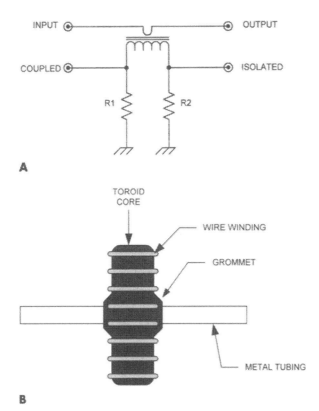

Figure 2.35
(A) Toroid core reflectometer;
(B) Toroid transformer detail.

Details of the pick-up sensor are shown in Figure 2.35B. The secondary is wound around the rim of the toroid in the normal manner, occupying not more than 330° of circumference. A rubber or plastic grommet is fitted into the center hole of the toroid core. The single-turn primary is formed by a single conductor passed once through the hole in the center of the grommet. It turns out the 3/16-in. o.d. (outside diameter) brass tubing (the kind sold in hobby shops that cater to model builders) will fit through several standard grommet sizes nicely, and will slip-fit over the center conductor of SO-239 coaxial connectors.

Another transmission line directional coupler is shown in Figure 2.36. Two short lengths of RG-58/U transmission line ($\approx$6-in.) are passed through a pair of toroid coils. Each coil is wound with 8 to 12 turns of wire. Note that the shields of the two transmission line segments are grounded only at one end.

Each combination of transmission line and toroid core form a transformer similar to the previous case. These two transformers are cross-coupled to form the network shown. The XMTR-ANTENNA path is the low-loss path, while (with the signal flow direction shown) the other two coupled ports are for forward and reflected power samples. These samples can be rectified and used to indicate the relative power levels flowing in the forward and reverse directions. Taken together these indications allow us to calculate VSWR.

Directional couplers are used for RF power sampling in measurement and transmitter control. They can also be used in receivers between either the mixer or the RF amplifier and the antenna input circuit. This arrangement can prevent the flow of LO signal and mixer products back towards the antenna, where they could be radiated and cause electromagnetic interference (EMI) to other devices.

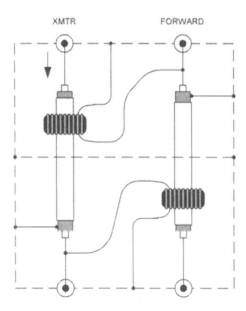

Figure 2.36
Transmission line directional coupler.

CHAPTER THREE

Smith Charting the Radio Frequency Circuit

The mathematics of transmission lines and other radio frequency (RF) devices becomes cumbersome at times, especially when dealing with complex impedances and "nonstandard" situations. In 1939, Phillip H. Smith published a graphical device for solving these problems; in 1945 he provided an improved version of the chart. That graphical aid, somewhat modified over time, is still in constant use in microwave electronics, as well as other fields where complex impedances and transmission line problems are found. The Smith chart is indeed a powerful tool for the RF designer.

The modern Smith chart is shown in Figure 3.1 and consists of a series of overlapping orthogonal circles (i.e., circles that intersect each other at right angles). This chapter will dissect the Smith chart, so that the origin and use of these circles is apparent. The set of orthogonal circles makes up the basic structure of the Smith chart.

THE NORMALIZED IMPEDANCE LINE

A baseline is highlighted in Figure 3.2 and bisects the Smith chart outer circle. This line is called the pure resistance line and forms the reference for measurements made on the chart. Recall that a complex impedance contains both resistance and reactance, and is expressed in the following mathematical form:

$$Z = R \pm jX \qquad\qquad [3.1]$$

Where:

Z is the complex impedance
R is the resistive component of the impedance
X is the reactive component of the impedance (may be capacitive, $-X_C$, or inductive, $+X_L$)

NAME	TITLE	DWG. NO
SMITH CHART FORM 82-BSPR (9-66)	KAY ELECTRIC COMPANY, PINE BROOK, N.J. © 1966 PRINTED IN U.S.A.	DATE

IMPEDANCE OR ADMITTANCE COORDINATES

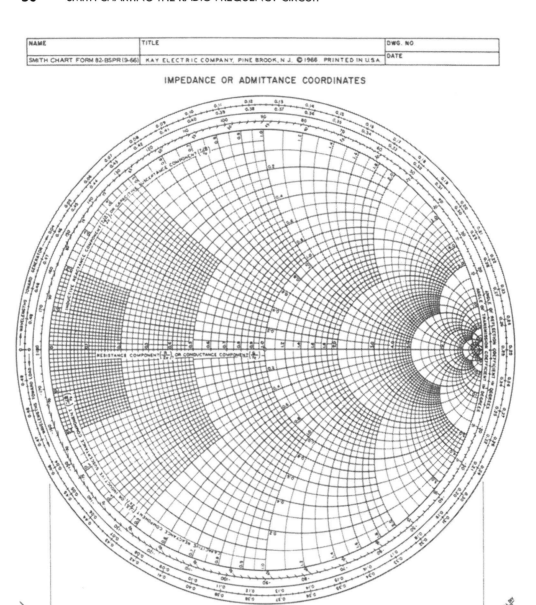

Figure 3.1 The Smith Chart.

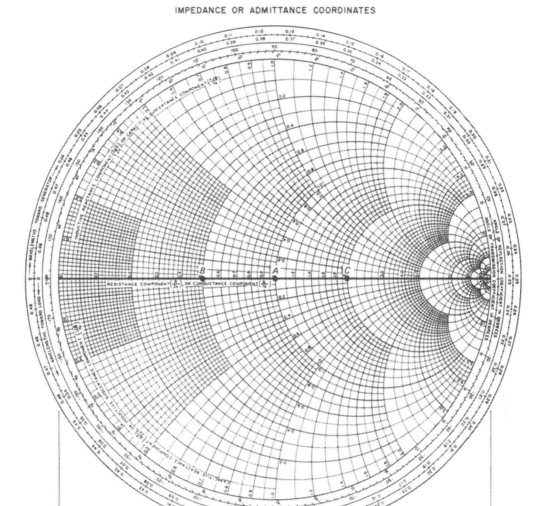

IMPEDANCE OR ADMITTANCE COORDINATES

Figure 3.2 Pure resistance line.

The *pure resistance line* represents the situation where $X = 0$, and the impedance is therefore equal to the resistive component only. In order to make the Smith chart universal, the impedances along the pure resistance line are *normalized* with reference to system impedance (e.g., Z_0 in transmission lines); for most microwave RF systems the system impedance is standardized at 50 Ω. In order to normalize the actual impedance, divide it by the system impedance. For example, if the load impedance of a transmission line is Z_L, and the characteristic impedance of the line is Z_0, then $Z = Z_L/Z_0$. In other words:

$$Z = \frac{R \pm jX}{Z_o} \qquad [3.1]$$

The pure resistance line is structured such that the system standard impedance is in the center of the chart and has a normalized value of 1.0 (see point "A" in Figure 3.2). This value derives from the fact that $Z_0/Z_0 = 1.0$.

To the left of the 1.0 point are decimal fraction values used to denote impedances less than the system impedance. For example, in a 50-Ω transmission line system with a 25-Ω load impedance, the normalized value of impedance is 25 Ω/50 Ω or 0.50 ("B" in Figure 3.2). Similarly, points to the right of 1.0 are greater than 1 and denote impedances that are higher than the system impedance. For example, in a 50-Ω system connected to a 100-Ω resistive load, the normalized impedance is 100 Ω/50 Ω, or 2.0; this value is shown as point "C" in Figure 3.2. By using normalized impedances, you can use the Smith chart for almost any practical combination of system—and load and/or source—impedances, whether resistive, reactive, or complex. Reconversion of the normalized impedance to actual impedance values is done by multiplying the normalized impedance by the system impedance. For example, if the resistive component of a normalized impedance is 0.45, then the actual impedance is:

$$Z = (Z_{normal})(Z_o) \qquad [3.3]$$

$$Z = (0.45)(50 \ \Omega) \qquad [3.4]$$

$$Z = 22.5 \ \Omega \qquad [3.5]$$

THE CONSTANT RESISTANCE CIRCLES

The *isoresistance circles*, also called the *constant resistance circles*, represent points of equal resistance. Several of these circles are shown highlighted in Figure 3.3. These circles are all tangential to the point at the right-hand extreme of the pure resistance line and are bisected by that line. When you construct complex impedances (for which X = nonzero) on the Smith chart, the points on these circles will all have the same resistive component. Circle "A," for example, passes through the center of the chart, so it has a normalized constant resistance of 1.0. Note that impedances that are pure resistances (i.e., $Z = R + j0$) will fall at the intersection of a constant resistance circle and the pure resistance line, and complex impedances (i.e., X not equal to zero) will appear at any other points on the circle. In Figure 3.2, circle "A" passes through the center of the chart, so it represents all points on the chart with a normalized resistance of 1.0. This particular circle is sometimes called the unity resistance circle.

THE CONSTANT REACTANCE CIRCLES

Constant reactance circles are highlighted in Figure 3.4. The circles (or circle segments) above the pure resistance line (Figure 3.4A) represent the inductive reactance (+X), and those circles (or segments) below the pure resistance line (Figure 3.4B) represent capacitive reactance (−X). In both cases, circle "A" represents a normalized reactance of 0.80. One of the outer circles (i.e., circle "A" in Figure 3.4C) is called the pure reactance circle.

Points along circle "A" represent reactance only: in other words, an impedance of $Z = 0 \pm jX$ (R = 0). Figure 3.4D shows how to plot impedance and admittance on the Smith chart. Consider an example in which system impedance Z_0 is 50 Ω, and the load impedance is $Z_L = 95 + j55$ Ω. This load impedance is normalized to:

$$Z = \frac{Z_L}{Z_o} \qquad [3.6]$$

$$Z = \frac{95 + j55\,\Omega}{50\,\Omega} \qquad [3.7]$$

$$Z = 1.9 + j1.1 \qquad [3.8]$$

An impedance radius is constructed by drawing a line from the point represented by the normalized load impedance, $1.9 + j1.1$, to the point represented by the normalized system impedance (1.0) in the center of the chart. A circle is constructed from this radius and is called the VSWR circle.

Admittance is the reciprocal of impedance, so it is found from:

$$Y = \frac{1}{Z} \qquad [3.9]$$

IMPEDANCE OR ADMITTANCE COORDINATES

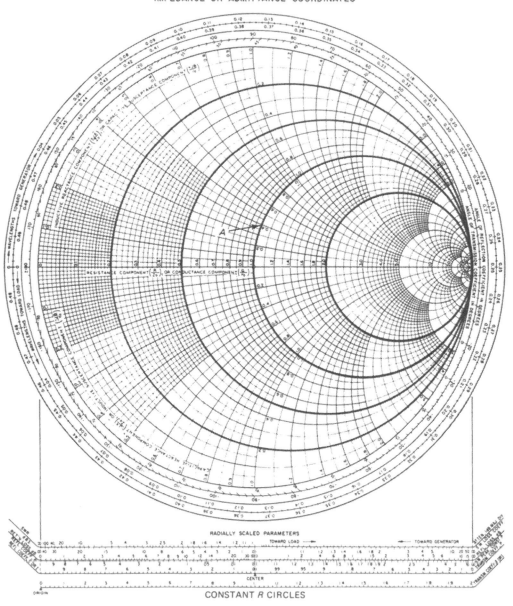

Figure 3.3 Isoresistance circles.

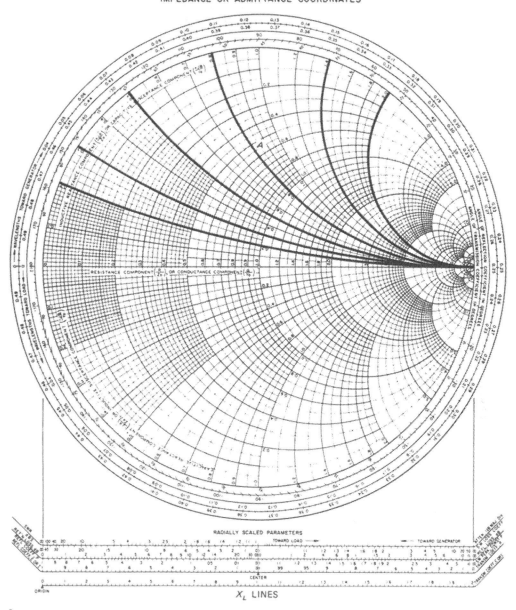

Figure 3.4 Constant reactance circles: (A) Inductive reactance; (B) Capacitive reactance; (C) Angle of transmission coefficient; (D) Plotting admittance and impedance on a Smith Chart.

Continued

Figure 3.4 *(continued)*

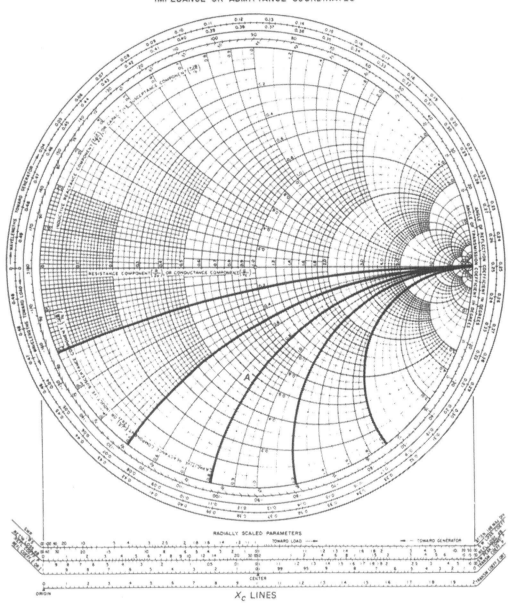

B

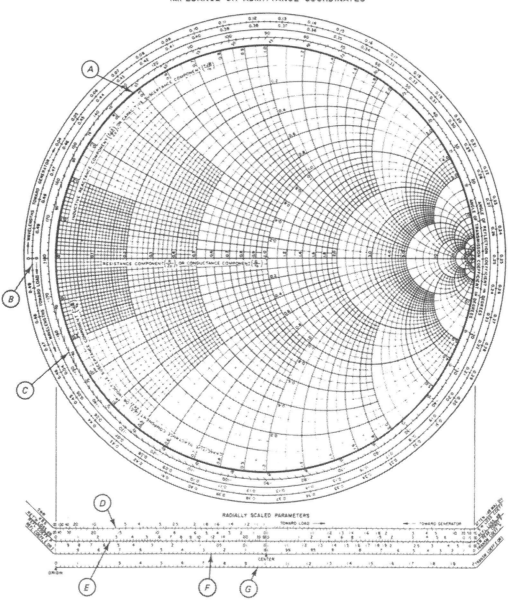

Continued

Figure 3.4 *(continued)*

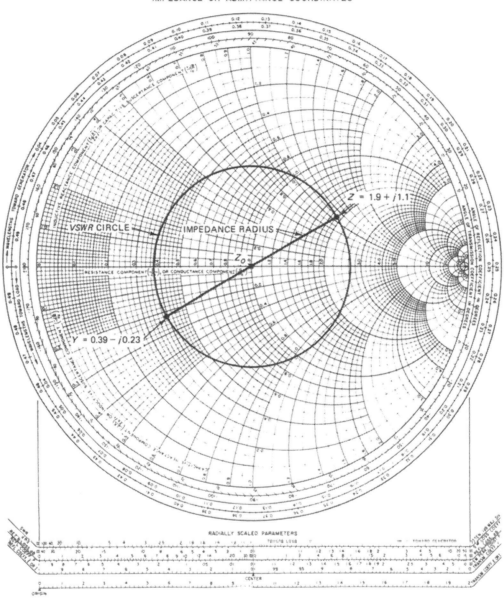

D

Because impedances in transmission lines are rarely pure resistive, but rather contain a reactive component as well, impedances are expressed using complex notation:

$$Z = R \pm jX \qquad\qquad [3.10]$$

Where:

Z is the complex impedance
R is the resistive component
X is the reactive component

In order to find the complex admittance, take the reciprocal of the complex impedance by multiplying the simple reciprocal by the complex conjugate of the impedance. For example, when the normalized impedance is $1.9 + j1.1$, the normalized admittance will be:

$$Y = \frac{1}{Z} \qquad\qquad [3.11]$$

$$Y = \left(\frac{1}{1.9 + j1.1}\right) - \left(\frac{1.9 - j1.1}{1.9 - j1.1}\right) \qquad\qquad [3.12]$$

$$Y = \frac{1.9 - j1.1}{3.6 + 1.2} \qquad\qquad [3.13]$$

$$Y = 0.39 - j0.23 \qquad\qquad [3.14]$$

One of the delights of the Smith chart is that this calculation is reduced to a quick graphical interpretation! Simply extend the impedance radius through the 1.0 center point until it intersects the VSWR circle again. This point of intersection represents the normalized admittance of the load.

OUTER CIRCLE PARAMETERS

The standard Smith chart shown in Figure 3.4C contains three concentric calibrated circles on the outer perimeter of the chart. Circle "A" has already been covered, and it is the pure reactance circle. The other two circles define the wavelength distance ("B") relative to either the load or generator end of the transmission line, and either the transmission, or reflection, coefficient angle in degrees ("C").

There are two scales on the wavelengths circle ("B" in Figure 3.4C), and both have their zero origin on the left-hand extreme of the pure resistance line. Both scales represent one-half wavelength for one entire revolution, and are calibrated from 0 through 0.50 such that these two points are identical with each other on the circle. In other words, starting at the zero point and traveling 360° around the circle brings one back to zero, which represents one-half wavelength, or 0.5 l.

Although both wavelength scales are of the same magnitude (0–0.50), they are opposite in direction. The outer scale is calibrated clockwise, and it represents wavelengths toward the generator; the inner scale is calibrated counterclockwise and represents wavelengths toward the load. These two scales are complementary at all points. Thus, 0.12 on the outer scale corresponds to (0.50–0.12) or 0.38 on the inner scale.

The angle of transmission coefficient and angle of reflection coefficient scales are shown in circle "C" in Figure 3.4C. These scales are the relative phase angle between reflected and incident waves. You may recall from transmission line theory that a short circuit at the load end of the line reflects the signal back toward the generator 180° out of phase with the incident signal; an open line (i.e., infinite impedance) reflects the signal back to the generator in-phase (i.e., 0°) with the incident signal. These facts are shown on the Smith chart by the fact that both scales start at 0° on the right-hand end of the pure resistance line, which corresponds to an infinite resistance, and it goes halfway around the circle to 180° at the 0 end of the pure resistance line. Note that the upper half-circle is calibrated 0 to +180°, and the bottom half-circle is calibrated 0 to +180°, reflecting indicative or capacitive reactance situations, respectively.

RADIALLY SCALED PARAMETERS

There are six scales laid out on five lines ("D" through "G" in Figure 3.4C and in expanded form in Figure 3.5) at the bottom of the Smith chart. These scales are called the radially scaled parameters—and they are both very important and often overlooked. With these scales, we can determine such factors as VSWR (both as a ratio and in decibels), return loss in decibels, voltage or current reflection coefficient, and the power reflection coefficient.

The reflection coefficient (Γ) is defined as the ratio of the reflected signal to the incident signal. For voltage or current:

$$\Gamma = \frac{E_{ref}}{E_{inc}} \qquad\qquad [3.15]$$

and

$$\Gamma = \frac{I_{ref}}{I_{inc}} \qquad\qquad [3.16]$$

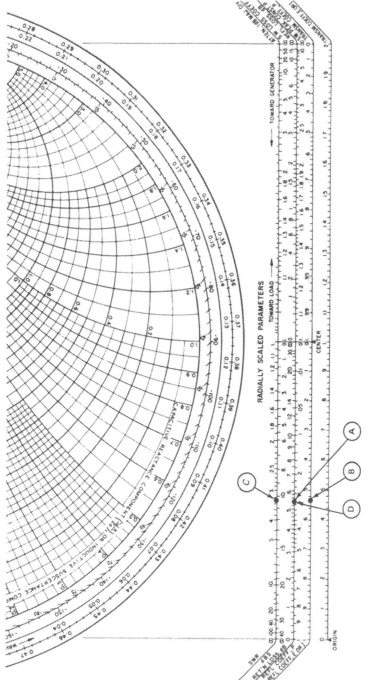

Figure 3.5 Radially scaled parameters on rule below the chart circles.

Power is proportional to the square of voltage or current, so:

$$P_{pwr} = \Gamma^2 \qquad\qquad [3.17]$$

or,

$$\Gamma = \frac{P_{ref}}{P_{inc}} \qquad\qquad [3.18]$$

EXAMPLE

Ten watts of microwave RF power is applied to a lossless transmission line, of which 2.8 watts is reflected from the mismatched load. Calculate the reflection coefficient.

$$\Gamma = \frac{P_{ref}}{P_{inc}} \qquad\qquad [3.19]$$

$$\Gamma_{pwr} = \frac{2.8\ watts}{10\ watts} \qquad\qquad [3.20]$$

$$\Gamma_{pwr} = 0.28 \qquad\qquad [3.21]$$

The voltage reflection coefficient (Γ) is found by taking the square root of the power reflection coefficient, so in this example it is equal to 0.529. These points are plotted at "A" and "B" in Figure 3.5.

Standing wave ratio (SWR) can be defined in terms of reflection coefficient:

$$VSWR = \frac{1 + \Gamma}{1 - \Gamma} \qquad\qquad [3.22]$$

or,

$$VSWR = \frac{1 + \sqrt{\Gamma_{pwr}}}{1 - \sqrt{\Gamma_{pwr}}} \qquad\qquad [3.23]$$

or, in our example:

$$VSWR = \frac{1\sqrt{0.28}}{1 - \sqrt{0.28}} \qquad [3.24]$$

$$VSWR = \frac{1 + 0.529}{1 - 0.529} \qquad [3.25]$$

or, in decibel form:

$$VSWR_{dB} = 20 \text{ LOG } (VSWR) \qquad [3.27]$$

$$VSWR_{dB} = 20 \text{ LOG } (20) \qquad [3.28]$$

$$VSWR_{dB} = (20)\,(0.510) = 10.2 \text{ dB} \qquad [3.29]$$

These points are plotted at "C" in Figure 3.5. Shortly, you will work on an example to show how these factors are calculated in a transmission line problem from a known complex load impedance.

Transmission loss is a measure of the one-way loss of power in a transmission line because of reflection from the load.

Return loss represents the two-way loss, so it is exactly twice the transmission loss. Return loss is found from:

$$\text{Loss}_{ret} = 10 \text{ LOG } (\Gamma_{pwr}) \qquad [3.30]$$

and, for our example in which $\Gamma_{pwr} = 0.28$:

$$\text{Loss}_{ret} = 10 \text{ LOG } (0.28) \qquad [3.31]$$

$$\text{Loss}_{ret} = (10)(-0.553) = -5.53 \text{ dB} \qquad [3.32]$$

This point is shown as "D" in Figure 3.5. The transmission loss coefficient can be calculated from:

$$TLC = \frac{1 + \Gamma_{pwr}}{1 - \Gamma_{pwr}} \qquad [3.33]$$

or, for our example:

$$TLC = \frac{1 + 0.28}{1 - 0.28} \qquad [3.34]$$

$$TLC = \frac{1.28}{0.72} = 1.78 \qquad [3.35]$$

The TLC is a correction factor that is used to calculate the attenuation caused by mismatched impedance in a "lossy," as opposed to the ideal "lossless," line. The TLC is found from laying out the impedance radius on the *Loss Coefficient scale* on the radially scaled parameters at the bottom of the chart.

SMITH CHART APPLICATIONS

One of the best ways to demonstrate the usefulness of the Smith chart is by practical example. The following sections look at two general cases: transmission line problems and stub matching systems.

Transmission Line Problems

Figure 3.6 shows a 50-Ω transmission line connected to a complex load impedance Z_L of $36 + j40\ \Omega$. The transmission line has a velocity factor (v) of 0.80, which means that the wave propagates along the line at 8-10 the speed of light (c = 300,000,000 m/s). The length of the transmission line is 28 cm. The generator (V_{in}) is operated at a frequency of 4.5 GHz and produces a power output of 1.5 watts. See what you can glean from the Smith chart (Figure 3.7).

First, normalize the load impedance. This is done by dividing the load impedance by the systems impedance (in this case, $Z_o = 50\ \Omega$):

$$Z = \frac{36 + j40\ \Omega}{50\ \Omega} \qquad [3.36]$$

$$Z = 0.72 + j0.8 \qquad [3.37]$$

The resistive component of impedance Z is located along the "0.72" pure resistance circle (see Figure 3.7). Similarly, the reactive component of impedance Z is located by traversing the 0.72 constant resistance circle until the +j0.8 constant reactance circle is intersected. This point graphically represents the normalized load impedance Z = 0.72 + j0.80. A VSWR circle is constructed with an impedance radius equal to the line between "1.0" (in the center of the chart) and the "0.72 + j0.8" point. At a frequency of 4.5 GHz, the length of a wave propagating in the transmission line, assuming a velocity factor of 0.80, is:

$$\lambda_{line} = \frac{c\,v}{F_{Hz}} \qquad [3.38]$$

$$\lambda_{line} = \frac{(3 \times 10^8\,m/s)(0.80)}{4.5 \times 10^9\,Hz} \qquad [3.39]$$

$$\lambda_{line} = \frac{2.4 \times 10^8\,m/s}{4.5 \times 10^9\,Hz} \qquad [3.40]$$

$$\lambda_{line} = 0.053\,m \times \frac{100\,cm}{m} = 5.3\,cm \qquad [3.41]$$

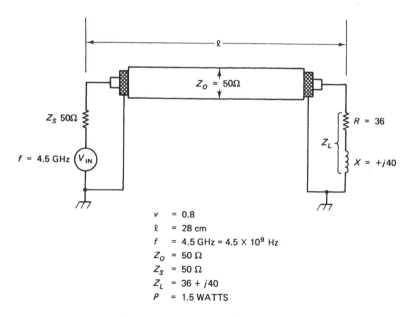

v = 0.8
ℓ = 28 cm
f = 4.5 GHz = 4.5 × 10⁹ Hz
Z_O = 50 Ω
Z_S = 50 Ω
Z_L = 36 + j40
P = 1.5 WATTS

Figure 3.6 Transmission line test circuit.

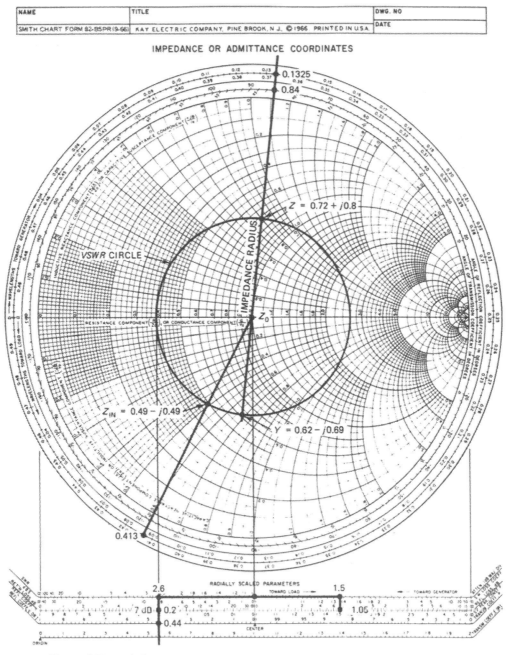

Figure 3.7 Results from test circuit.

One wavelength is 5.3 cm, so a half wavelength is 5.3 cm/2, or 2.65 cm. The 28 cm line is 28 cm/5.3 cm, or 5.28 wavelengths long. A line drawn from the center (1.0) to the load impedance is extended to the outer circle, and it intersects the circle at 0.1325. Because one complete revolution around this circle represents one-half wavelength, 5.28 wavelengths from this point represent ten revolutions plus 0.28 more. The residual 0.28 wavelengths are added to 0.1325 to form a value of (0.1325 + 0.28), or 0.413. The point "0.413" is located on the circle and is marked. A line is then drawn from 0.413 to the center of the circle, and it intersects the VSWR circle at $0.49 - j0.49$, which represents the input impedance (Z_{in}) looking into the line. To find the actual impedance represented by the normalized input impedance, you have to "denormalize" the Smith chart impedance by multiplying the result by Z_o:

$$Z_{in} = (0.49 - j0.49)(50 \ \Omega) \qquad \text{[3.42]}$$

$$Z_{in} = 24.5 - j24.5 \ \Omega \qquad \text{[3.43]}$$

It is this impedance that must be matched at the generator by a conjugate-matching network. The admittance represented by the load impedance is the reciprocal of the load impedance; it is found by extending the impedance radius through the center of the VSWR circle until it intersects the circle again. This point is found, representing the admittance $Y = 0.62 - j0.69$. Confirm the solution mathematically as follows:

$$Y = \frac{1}{Z} \qquad \text{[3.44]}$$

$$Y = \frac{1}{0.72 + j0.80} - \frac{0.72 - j0.80}{0.72 - j0.80} \qquad \text{[3.45]}$$

$$Y = \frac{0.72 + j0.80}{1.16} = 0.62 - j0.69 \qquad \text{[3.46]}$$

The VSWR if found by transferring the "impedance radius" of the VSWR circle to the radial scales below. The radius (0.72 − 0.80) is laid out on the VSWR scale (topmost of the radially scaled parameters) with a pair of dividers from the center mark, and we find that the VSWR is approximately 2.6:1. The decibel form of VSWR is 8.3 dB (next scale down from VSWR), and this is confirmed by:

$$VSWR_{dB} = 20 \ LOG \ (VSWR) \qquad\qquad [3.47]$$

$$VSWR_{dB} = (20) \ LOG \ (2.7) \qquad\qquad [3.48]$$

$$VSWR_{dB} = (20)(0.431) = 8.3 \ dB \qquad\qquad [3.49]$$

The transmission loss coefficient is found in a manner similar to the VSWR, using the radially scaled parameter scales. In practice, once you have found the VSWR you need only drop a perpendicular line from the 2.6:1 VSWR line across the other scales. In this case, the line intersects the voltage reflection coefficient at 0.44. The power reflection coefficient (G_{pwr}) is found from the scale, and is equal to G2. The perpendicular line intersects the power reflection coefficient line at 0.20. The angle of reflection coefficient is found from the outer circles of the Smith chart. The line from the center to the load impedance ($Z = 0.72 + j0.80$) is extended to the Angle of Reflection Coefficient in Degrees circle, and intersects it at approximately 84°. The reflection coefficient is therefore 0.44/84°. The transmission loss coefficient (TLC) is also found from the radially scaled parameter scales. In this case, the impedance radius is laid out on the Loss Coefficient scale, where it is found to be 1.5. This value is confirmed from:

$$TLC = \frac{1 + \Gamma_{pwr}}{1 - \Gamma_{pwr}} \qquad\qquad [3.50]$$

$$TLC = \frac{1 + 0.20}{1 - 0.21} \qquad\qquad [3.51]$$

$$TLC = (1.20/0.79) = 1.5 \qquad\qquad [3.52]$$

The Return Loss is also found by dropping the perpendicular from the VSWR point to the RETURN LOSS, dB line, and the value is found to be approximately 7 dB, which is confirmed by:

$$Loss_{ret} = 10 \ LOG \ (G_{pwr}) \ dB \qquad\qquad [3.53]$$

$$\text{Loss}_{\text{ret}} = 10 \text{ LOG } (0.21) \text{ dB} \tag{3.54}$$

$$\text{Loss}_{\text{ret}} = (10)(-0.677) \text{ dB} \tag{3.55}$$

$$\text{Loss}_{\text{ret}} = 6.77 \text{ dB} = -6.9897 \text{ dB} \tag{3.56}$$

The reflection loss is the amount of RF power reflected back down the transmission line from the load. The difference between incident power supplied by the generator (1.5 watts in this example), $P_{\text{inc}} - P_{\text{ref}} = P_{\text{abs}}$ and the reflected power, is the absorbed power (P_a); or in the case of an antenna, the radiated power. The reflection loss is found graphically by dropping a perpendicular from the TLC point (or by laying out the impedance radius on the REFL. Loss, dB scale); in this example (Figure 3.7) it is −1.05 dB. You can check the calculations. The return loss was −7 dB, so:

$$-7 \, dB = 10 \text{ LOG}\left(\frac{P_{ref}}{P_{inc}}\right) \tag{3.57}$$

$$-7 = 10 \text{ LOG}\left(\frac{P_{ref}}{1.5 \, watts}\right) \tag{3.58}$$

$$\frac{-7}{10} = \text{LOG}\left(\frac{P_{ref}}{1.5 \, watts}\right) \tag{3.59}$$

$$10^{(-7/10)} = \frac{P_{ref}}{1.5 \, watts} \tag{3.60}$$

$$0.2 = \frac{P_{ref}}{1.5 \, watts} \tag{3.61}$$

$$(0.2)(1.5 \, watts) = P_{ref} \tag{3.62}$$

$$0.3 \text{ watts} = P_{\text{ref}} \tag{3.63}$$

The power absorbed by the load (P_a) is the difference between incident power (P_{inc}) and reflected power (P_{ref}). If 0.3 watts is reflected, then that means the absorbed power is (1.5 − 0.3), or 1.2 watts. The reflection loss is −1.05 dB and can be checked from:

$$-1.05 \, dB = 10 \, \mathrm{LOG}\left(\frac{P_a}{P_{inc}}\right) \qquad [3.64]$$

$$\frac{-1.05 \, dB}{10} = \frac{P_a}{1.5 \, watts} \qquad [3.65]$$

$$10^{(-1.05/10)} = \frac{P_a}{1.5 \, watts} \qquad [3.66]$$

$$0.785 = \frac{P_a}{1.5 \, watts} \qquad [3.67]$$

$$(1.5 \text{ watts}) \times (0.785) = P_a \qquad [3.68]$$

$$1.2 \text{ watts} = P_a \qquad [3.69]$$

Now check what you have learned from the Smith chart. Recall that 1.5 watts of 4.5-GHz microwave RF signal were input to a 50-Ω transmission line that was 28 cm long. The load connected to the transmission line has an impedance of 36 + j40. From the Smith chart:

Admittance (load):	0.62 − j0.69
VSWR:	2.6:1
VSWR(dB):	8.3 dB
Refl. coef. (E):	0.44
Refl. coef. (P):	0.2
Refl. coef. angle:	84°
Return loss:	−7 dB
Refl. loss:	−1.05 dB
Trans. loss. coef.:	1.5

Note that in all cases the mathematical interpretation corresponds to the graphical interpretation of the problem, within the limits of accuracy of the graphical method.

Stub Matching Systems

A properly designed matching system will provide a conjugate match to a complex impedance. Some sort of matching system or network is needed any time the load impedance (Z_L) is not equal to the characteristic impedance (Z_o) of the transmission line. In a transmission line system, it is possible to use a shorted stub connected in parallel with the line, at a critical distance back from the mismatched load, in order to affect a match. The stub is merely a section of transmission line that is shorted at the end not connected to the main transmission line. The reactance (hence also susceptance) of a shorted line can vary from −1 to +1, depending upon length, so you can use a line of critical length L2 to cancel the reactive component of the load impedance. Because the stub is connected in parallel with the line it is a bit easier to work with admittance parameters, rather than impedance.

Consider the example of Figure 3.8, in which the load impedance is $Z = 100 + j60$, which is normalized to $2.0 + j1.2$. This impedance is plotted on the Smith chart in Figure 3.9, and a VSWR circle is constructed. The admittance is found on the chart at point $Y = 0.37 - j0.22$.

In order to provide a properly designed matching stub, you need to find two lengths. L1 is the length (relative to wavelength) from the load toward the generator (see L1 in Figure 3.8); L2 is the length of the stub itself.

The first step in finding a solution to the problem is to find the points where the unit conductance line (1.0 at the chart center) intersects the VSWR circle; there are two such points shown in Figure 3.9: $1.0 + j1.1$ and $1.0 - j1.1$. Select one of these (choose $1.0 + j1.1$) and extend a line from the center 1.0 point through the $1.0 + j1.1$ point to the outer circle (WAVELENGTHS TOWARD GENERATOR). Similarly, a

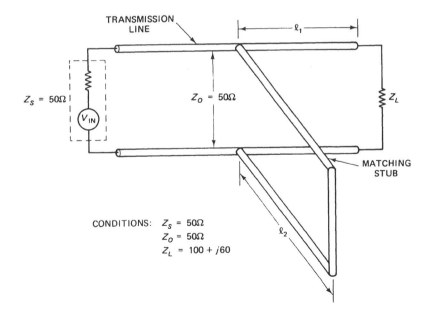

Figure 3.8 Matching stub on transmission line.

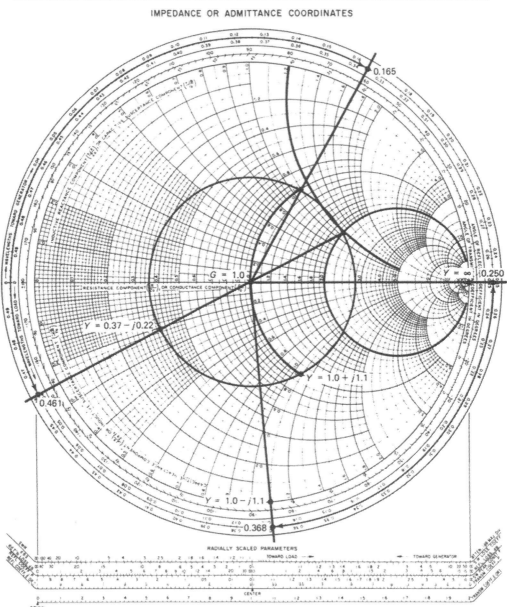

Figure 3.9 Smith Chart solution to problem.

line is drawn from the center through the admittance point $0.37 - 0.22$ to the outer circle. These two lines intersect the outer circle at the points 0.165 and 0.461. The distance of the stub back toward the generator is found from:

$$L1 = 0.165 + (0.500 - 0.461)\lambda \qquad [3.70]$$

$$L1 = 0.165 + 0.039\,\lambda \qquad [3.71]$$

$$L1 = 0.204\,\lambda \qquad [3.72]$$

The next step is to find the length of the stub required. This is done by finding two points on the Smith chart. First, locate the point where admittance is infinite (far right side of the pure conductance line); second, locate the point where the admittance is $0 - j1.1$ (note that the susceptance portion is the same as that found where the unit conductance circle crossed the VSWR circle). Because the conductance component of this new point is 0, the point will lay on the $-j1.1$ circle at the intersection with the outer circle. Now draw lines from the center of the chart through each of these points to the outer circle. These lines intersect the outer circle at 0.368 and 0.250. The length of the stub is found from:

$$L2 = (0.368 - 0.250)\lambda \qquad [3.73]$$

$$L2 = 0.118\,\lambda \qquad [3.74]$$

From this analysis you can see that the impedance, $Z = 100 + j60$, can be matched by placing a stub of a length 0.118 l at a distance 0.204 l back from the load.

THE SMITH CHART IN LOSSY CIRCUITS

Thus far, you have dealt with situations in which loss is either zero (i.e., ideal transmission lines), or so small as to be negligible. In situations where there is appreciable loss in the circuit or line, however, you see a slightly modified situation. The VSWR circle, in that case, is actually a spiral, rather than a circle.

Figure 3.10 shows a typical situation. Assume that the transmission line is 0.60l long and is connected to a normalized load impedance of $Z = 1.2 + j1.2$. An "ideal" VSWR circle is constructed on the impedance radius represented by $1.2 + j1.2$. A line ("A") is drawn, from the point where this circle intersects the pure resistance baseline ("B"), perpendicularly to the ATTEN 1 dB/MAJ. DIV. line on the radially scaled parameters. A distance representing the loss (3 dB) is stepped off on this scale. A second perpendicular line is drawn, from the −3 dB point, back to the pure resistance line ("C"). The point where line "C" intersects the pure resistance line becomes the radius for a new circle that contains the actual input impedance of the line. The length of the line is 0.60l, so you must step back (0.60 − 0.50)l or 0.1l. This point is located on the WAVELENGTHS TOWARD GENERATOR outer circle. A line is drawn from this point to the 1.0 center point. The point where this new line intersects the new circle is the actual input impedance (Z_{in}). The intersection occurs at 0.76 + j0.4, which (when denormalized) represents an input impedance of 38 + j20 ohms.

FREQUENCY ON THE SMITH CHART

A complex network may contain resistive, inductive reactance, and capacitive reactance components. Because the reactance component of such impedances is a function of frequency, the network or component tends also to be frequency sensitive. You can use the Smith chart to plot the performance of such a network with respect to various frequencies. Consider the load impedance connected to a 50-Ω transmission line in Figure 3.11. In this case, the resistance is in series with a 2.2-pF capacitor, which will exhibit a different reactance at each frequency. The impedance of this network is:

$$Z = R - j\left(\frac{1}{\lambda C}\right) \qquad [3.75]$$

or,

$$Z = 50 - j\left(\frac{1}{2\pi F C}\right) \qquad [3.76]$$

And, in normalized form:

$$Z' = 1.0 - \frac{j}{(2\pi F C) \times 50} \qquad [3.77]$$

$$Z' = 1.0 - \frac{j}{6.9 \times 10^{-10} F} \qquad [3.78]$$

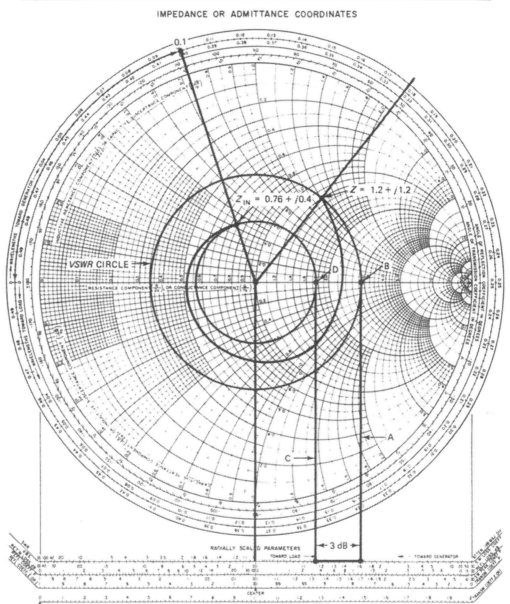

IMPEDANCE OR ADMITTANCE COORDINATES

Figure 3.10 Smith Chart solution for a lossy line.

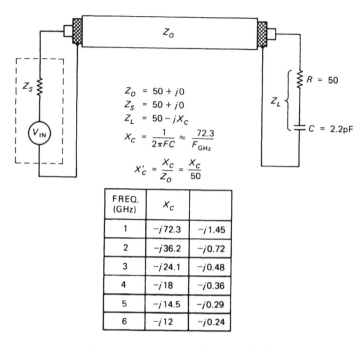

Figure 3.11   Test circuit for reactive load.

or, converted to GHz:

$$Z' = 1.0 - \frac{j72.3}{F_{GHz}} \qquad [3.80]$$

The normalized impedances for the sweep of frequencies from 1 to 6 GHz is, therefore:

$$Z = 1.0 - j1.45 \qquad [3.81]$$

$$Z = 1.0 - j0.72 \qquad [3.82]$$

$$Z = 1.0 - j0.48 \qquad [3.83]$$

$$Z = 1.0 - j0.36 \qquad\qquad\qquad\qquad [3.84]$$

$$Z = 1.0 - j0.29 \qquad\qquad\qquad\qquad [3.85]$$

$$Z = 1.0 - j0.24 \qquad\qquad\qquad\qquad [3.86]$$

These points are plotted on the Smith chart in Figure 3.12. For complex networks, in which both inductive and capacitive reactance exist, take the difference between the two reactances (i.e., $X = X_L - X_C$).

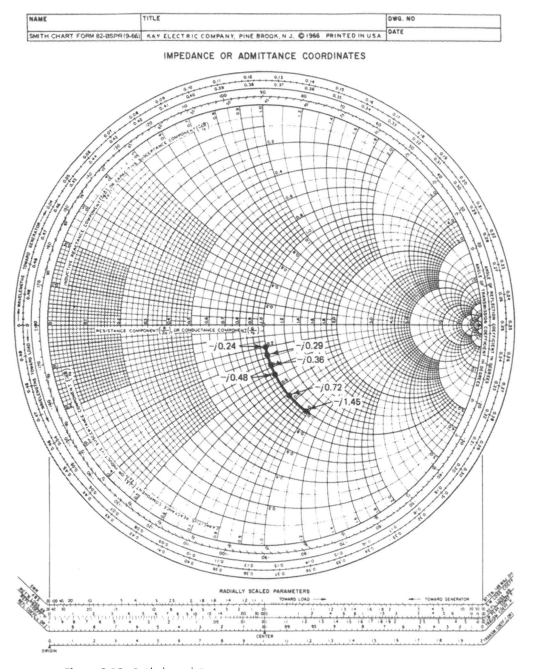

Figure 3.12   Smith chart solution.

CHAPTER FOUR

Signal Sources and Signal Generators

Signal generators and signal sources are instruments that generate controlled signals for use in testing and measurement. There is a distinction made by some people between *signal sources* and *signal generators.* The former produce continuous wave (CW) output signals without modulation, while the latter will produce one or more forms of modulated signal (AM, FM, SSB, PM) in addition to CW output. In many cases, however, you will see the words "source" and "generator" used interchangeably in popular usage.

Some signal sources produce a single output frequency (or a discrete number of fixed output frequencies). These instruments are sometimes used for testing channelized receiver systems. Other signal sources will produce outputs over a very wide range of frequencies. Add a modulator stage to these instruments and a signal generator is created. Figure 4.1 shows several typical commercial signal generators and sources.

GRADES OF INSTRUMENT

Signal generators and sources come in several grades. Which one to select depends on the use. A *service grade* instrument (Figure 4.1D) is used for troubleshooting common broadcast band receivers. They often lack a calibrated or metered output level control, and the frequency accuracy is usually low. More importantly for many sensitive measurements (e.g., see Chapter 8, "Radio Receivers and Their Measurements") at low output levels, more signal will escape around the flanges than comes through the output connector. Such instruments are useful for simple troubleshooting, but useless for accurate measurements.

Laboratory grade signal sources and generators are very-high-quality instruments with accurate frequency readout and output level controls. These instruments are used in making lab measurements of receivers and other devices where high precision and accuracy is desired.

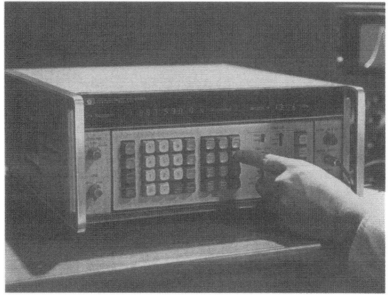

A

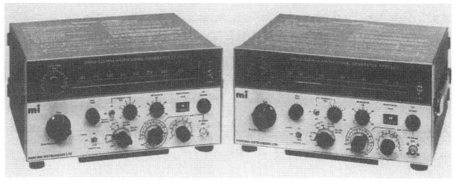

B

Figure 4.1 Typical signal generators: (A) Digital readout synthesizer; (B) Analog readout signal generators; (C) Sweep generator; (D) Service-grade signal generator.

Quality service grade instruments fall somewhere between the two previous grades. Several mainline manufacturers of high-quality laboratory signal sources and generators also manufacture "economy" lines that fit this category. They are considerably higher grade than the simple service instruments, but are not up to the lab grade. They are often used for troubleshooting high-quality telecommunications and landmobile systems where the highly accurate and precise measurements are not needed.

C

D

Output Level

The output of a calibrated signal generator is usually expressed in either *microvolts* (μV) or *dBm* (decibels relative to 1 milliwatt in 50 ohms), or both μV and dBm. It is useful to have a feel for both forms of output level indication. One-microvolt

(1 μV) is 10^{-6} volts, so when applied across a 50-ohm resistive load produces a power level of

$$V^2/R = (10^{-6}\,V)^2/(50\,\Omega) = 2 \times 10^{-14}\,\text{watts} \qquad [4.1]$$

The 0 dBm reference level is one-milliwatt (1 mW) dissipated in a 50-ohm resistive load. This represents an applied voltage of

$$V = \sqrt{P\,R} = \sqrt{(0.001W)(50\,\Omega)} = 0.2236\,volts \qquad [4.2]$$

If the output level set dial on a particular instrument is not calibrated in the correct units, then the required unit can be calculated using these methods.

To find the output power level in watts or milliwatts from dBm is similarly simple:

$$P = 10^{dBm/10}\,\text{milliwatts} \qquad [4.3]$$

To find the level in watts divide Equation 4.3 by 1000.

Output Signal Quality

It would be nice if all signal sources and signal generators were ideal, that is, that the output frequency and output level were noiseless and perfectly calibrated. That never occurs, although the differences in these specifications are a principal difference between high- and low-quality instruments.

Output Frequency

The important considerations regarding frequency are the *range, resolution, accuracy,* and (in automatic test equipment applications) the *switching speed.*

Range

The frequency range is a specification that delineates the specific frequencies that are covered. In some cases, there will be only one frequency, or some small number of discrete frequencies. In other cases, one or more bands of frequencies are provided.

Resolution

The resolution is the statement of the smallest increment of frequency that can be set. On analog instruments that do not have a counter the resolution is poor. The resolution may (but not certainly) be improved by adding a digital frequency counter to measure the output frequency. On modern synthesizers it is possible to set frequency with extremely good resolution.

Accuracy

This specification refers to how nearly the actual output frequency matches the set frequency. The accuracy is a function of the set frequency (and how closely it can be set), F_{Set}, long-term aging (τ_{aging}), and the time since last calibration (τ_{cal}). Mathematically, these are represented as follows:

$$\text{Accuracy} = \pm\ F_{Set} \times \tau_{Aging} \times \tau_{Cal} \qquad [4.4]$$

EXAMPLE

A signal generator is set to 480 MHz and has an aging rate of 0.155 ppm/year. It has been six months (0.5 year) since the last calibration.

$$\text{Accuracy} = \pm\ F_{Set} \times \tau_{Aging} \times \tau_{Cal}$$

$$\text{Accuracy} = \pm\ (480\ \text{MHz}) \times (0.155\ \text{ppm/year}) \times (0.5\ \text{year})$$

$$\text{Accuracy} = \pm\ 37.2\ \text{Hz}$$

There may also be some random variation in the output frequency. Figure 4.2 shows the uncertainty band around the set frequency. The actual output frequency, F_o, will be $F_{Set} \pm Accuracy$. It is the general practice to calibrate a signal generator on six-month or annual schedules, depending on the use.

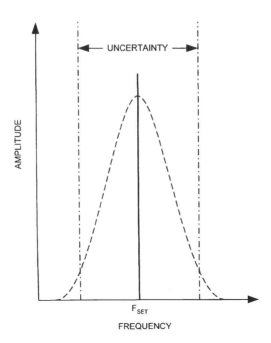

Figure 4.2
Frequency uncertainty.

Switching Speed (Settling Time)

This is the length of time, usually in milliseconds or microseconds, that is required for a synthesized signal source or generator to move to a new frequency when digitally commanded to change. It is calculated as the length of time for the error of the frequency and/or output level commanded by the change to come into specification range.

Output Level

The output level can be expressed in either voltage, power, or dBm notations. All are equivalent, although one or the other will be preferred in most cases. The most common method of describing the output level is dBm. A typical signal generator or CW signal source will produce output levels from −136 dBm to +10 dbm (some go to higher levels), with an output level accuracy of ±0.50 dBm and a resolution of 0.01 to 0.02 dB.

As with frequency there are factors that affect the accuracy of the actual output versus the set output. Figure 4.3 shows that there is a zone of uncertainty around the output level set. For any given setting there are P_{min} and P_{Max} values. For example, if the only error is the accuracy discussed above (e.g., 0.50 dB), a level set of, say, −10 dBm will produce an actual output power level of −9.5 to −10.5 dBm.

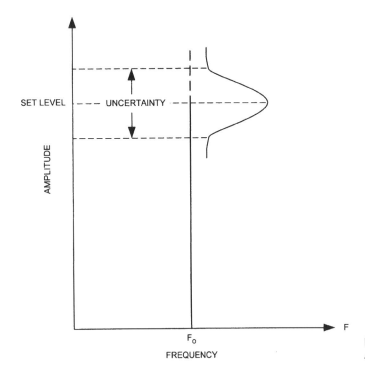

Figure 4.3
Amplitude uncertainty.

Spectral Purity

The output signal is not always nice and clean. Although the purity of the output signal is one of the distinguishing factors that differentiate lower-quality and higher-quality generators, they all produce signals other than the one desired. Figure 4.4A shows a typical spectrum output. This display is what might be seen on a spectrum analyzer. The main signal is a CW sine wave, so ideally we would expect only one single spike, with a height proportional to the output level. But there are a lot of other signals present.

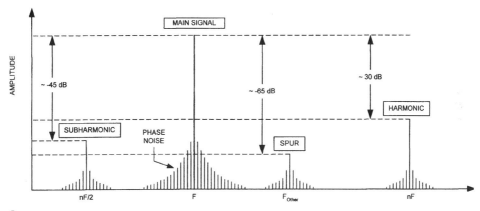

A

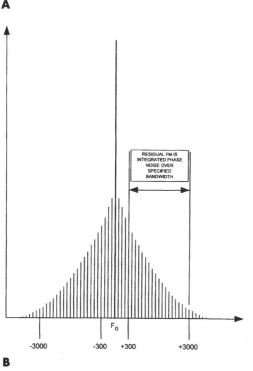

B

Figure 4.4
(A) Output spectrum of a signal source; (B) Phase noise and residual FM spectra.

First, note that the main signal is spread out by *phase noise*. This noise is random variation around the main frequency. When integrated over a specified bandwidth, such as 300 to 3,000 Hz, the phase noise is called *residual FM* (Figure 4.4B).

Second, there are harmonics present. If the main signal has a frequency of F, the harmonics have frequencies of nF, where n is an integer. For example, the second harmonic is 2F, and the third harmonic is 3F. In many cases, the 3F harmonic is stronger than the 2F harmonic, although in general the higher harmonics are weaker than lower harmonics.

There are also sometimes subharmonics. These are integer quotients of the main signal. Again, if the F is the main signal frequency, nF/2 represents the subharmonics. Typically, unless something is interfering with the output signal, subharmonics are not as prominent. One thing that does make subharmonics prominent, however, is the use of frequency multiplier or divider stages (which is the case in many modern generators).

Finally, there are miscellaneous spurious signals ("spurs") found on some generators. These might be due to power supply ripples that modulate the output signal; parasitic oscillations; digital noise from counter- or phase-locked loop circuits; and other sources.

Harmonics and spurs are usually measured in terms of *decibels below the carrier* (dBc), where the carrier is the amplitude of the main output signal. In general, the lower the unwanted components are, the better the signal source.

Phase noise warrants some special consideration. It is usually measured in terms of dBc/Hz, or decibels below the carrier per hertz of bandwidth. This noise is concentrated around the main signal frequency; it is normally graphed on a log-log scale to permit both close-in and further-out noise components to be compressed on one graph.

ARCHITECTURES

Although there are many different configurations, with different "block diagram" representations, there are only a few different architectures used in designing signal generators. Figure 4.5 shows a simple analog architecture that was once common on even high-grade instruments, and is still common on service grade instruments.

The signal is generated in an L-C controlled *variable frequency oscillator* (VFO). The VFO typically has a bandswitch for selecting different frequency ranges. A calibrated tuning dial gives the user an approximate idea of the output frequency. However, because of drift and the mechanical aspects of calibrating the dial, these dials are not terribly accurate.

Some instruments have an output amplifier, although for many decades even quality signal generators lacked power amplifiers. The output of the VFO was fed directly to the output level control.

Service grade generators of this architecture usually have a crude form of output level control. Higher quality instruments, on the other hand, will have some variant of the output circuit shown in Figure 4.5. A *high-level output* is sometimes

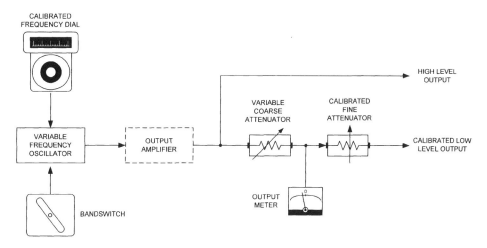

Figure 4.5 Block diagram of a simple signal generator.

provided to permit the user to route the signal to a frequency counter so that an accurate determination of frequency can be achieved.

There are two attenuators in the output-level setting circuit. A coarse attenuator is used to set a relative output meter to some calibrated point. In most cases, the meter would be calibrated with a zero in the center of the analog scale. The coarse attenuator is adjusted to center the meter pointer over the zero point in the center of the meter. When this is done, the settings of the fine output attenuator are valid.

A *synthesizer* architecture is shown in Figure 4.6. This type of signal generator is more modern and is capable of producing very accurate, high-quality signals. There are three main sections to this signal source: *Reference Section, Frequency Synthesizer,* and *Output Section.* In Figure 4.6 each section is broken down into further components for the sake of easy analysis.

Reference Section

The Reference Section is at the very core of the signal generation process. It is an accurate, stable, fixed frequency source such as a crystal oscillator. The frequency of the reference section must be precisely adjustable over a small range so that it can be compared to a higher-order standard, such as a Cesium beam oscillator or WWVB comparator receiver, for purposes of calibration.

Because it controls the frequency synthesizer, the stability of the reference section determines the overall stability of the signal generator. The stability of ordinary crystal oscillators is reasonably good for many purposes, but not for use in the reference section of a signal source. For that purpose, either *Temperature Compensated Crystal Oscillators* (TCXO) or *Oven Controlled Crystal Oscillators* (OCXO) are used for the reference section. The TCXO will typically exhibit crystal aging of better then ± 2 ppm/year, and a temperature aging of ± 1 ppm/year. The OCXO is capable of 0.1 ppm/year for aging and 0.01 ppm/year for temperature.

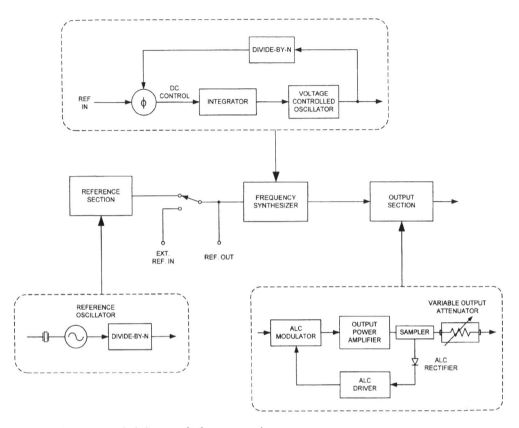

Figure 4.6 Block diagram of a frequency synthesizer.

The crystal oscillator is usually operated at a frequency such as 5 MHz, but lower frequencies are often needed. To generate these lower frequencies a divide-by-N digital counter is provided. This circuit will divide the output frequency by some integer, N, to produce a much lower reference frequency.

Many signal generators provide *REF. OUT* and *EXT. REF. IN* capability (these connections are often located on the rear panel of signal generators, so they may be overlooked). The *REF. OUT* connector provides the reference signal to other instruments, or it can be used for calibration purposes. The *EXT. REF. IN* allows the use of an external reference source in place of the internal reference section. This is done sometimes to lock up two signal generators that must work together, or to substitute a much more accurate reference such as a Cesium beam clock.

Frequency Synthesizer

The actual signal is produced in the frequency synthesizer section. It is generated by a *voltage controlled oscillator* (VCO) whose output is compared to the reference signal. Voltage variable capacitance diode ("varactors") can be used for the VCO, as can *surface acoustical wave* (SAW) oscillators (which are used at higher frequencies and in the microwave bands).

The frequency of the VCO is set by a DC control voltage applied to its tuning input line. This control voltage is generated by integrating the output of a phase detector or phase comparator that receives the reference frequency, with a divide-by-N version of the VCO frequency as inputs. When the two frequencies are equal, then the output of the phase detector is zero, so the VCO tuning voltage is at some quiescent value.

But if the frequency of the VCO drifts, the phase detector output becomes nonzero. The integrator output ramps up in the direction that will cancel the frequency drift of the VCO. The VCO frequency is continuously held in check by corrections from the integrated output of the phase detector. This type of circuit is called a *phase-locked loop* (PLL).

Suppose, for example, that a signal source has a reference of 5 MHz, and it is divided by 20 to produce a 250 kHz reference. If the frequency synthesizer divide-by-N stage is set for, say, $N = 511$, then the VCO output frequency will be $0.25 \text{ MHz} \times 511 = 127.75 \text{ MHz}$. Band switching, operating frequency, and frequency resolution are controlled by manipulating the reference frequency divide-by-N and VCO divide-by-N settings. In some cases, the frequency is entered by keypad, and this tells the signal generator how to set these values. Alternatively, "tunable" signal generators may have a digital encoder shaft connected to a front panel control.

Noise Floor

The noise component of the output signal is composed of *thermal noise* and *phase noise*. Of these two, the phase noise tends to dominate the performance of the signal source. Both the Reference Oscillator and VCO phase noise contributes to the overall output phase noise. There is a 20 LOG(N) degradation of the phase noise performance of the signal source because of the divide-by-N nature of the PLL. Fortunately, the bandwidth of the PLL tends to limit the contribution of the VCO phase noise to the overall phase noise performance.

Output Section

The output section performs three basic functions: it boosts power output to a specified maximum level, it provides precision control over the actual output level, and it keeps the output level constant as frequency is changed.

The power amplifier is a wideband amplifier that produces an output level of some value in excess of the required maximum output level (e.g., +13 dBm). A calibrated precision attenuator can then be used to set the actual output level to any lower value required (e.g., −136 dBm to +13 dBm).

The accuracy of the output power setting is dependent on keeping the RF power applied to the attenuator input constant, despite the fact that oscillators (including VCOs) tend to exhibit output signal amplitude changes as frequency is changed. In older manual signal sources there was a zero-center RF voltmeter at the input of the attenuator, and the output coarse attenuator level could be manually set to the zero line, making the calibration of the fine attenuator meaningful.

Modern signal sources, however, use an *automatic level control* (ALC) circuit to accomplish this job.

Automatic Level Control (ALC)

The ALC modulator is essentially an amplitude modulator that is controlled by a DC voltage developed by rectifying and filtering a sample of the RF output level. The ALC driver compares the actual output level with a preset value, and adjusts the control signal to the ALC modulator in a direction that cancels the error.

MODULATORS

A signal generator is said to differ from a signal source in that it will provide modulation of the CW signal. Although I believe that is a distinction without a practical difference, it is nonetheless a commonly held usage.

Amplitude Modulation (AM)

Amplitude modulation conveys intelligence over a radio carrier by means of varying the amplitude in accordance with the applied audio. The sine wave RF carrier is described by a(t) = A_c Sin(2 π f_c t), where a(t) is the instantaneous amplitude, A_c is the peak amplifier, f_c is the carrier frequency, and t is time. When another signal is used to vary the amplitude of the carrier, amplitude modulation (AM) results:

$$s(t) = A_c\, Sin\,(2\pi f_c\, t) \times [1 + \mu Sin(2\pi f_m\, t)] \qquad [4.5]$$

Where

μ is the depth of modulation
f_m is the modulating frequency
(all other terms previously defined)

You will see the terms *linear AM* and *logarithmic AM*. The former is used when the depth of modulation is expressed as a percentage, and the latter when it is expressed in decibels (dB). Figure 4.7 shows the block diagram of the output section of a signal generator that offers amplitude modulation. The depth of modulation for most signal generators is adjustable over at least 0 to 30%, while some offer 100% AM.

The amplitude modulator is called a burst modulator in some signal generators and is placed between the ALC modulator and the output power amplifier stage. The input signal used to modulate the RF carrier can be derived from an external source, or provided by an internal function generator that produces at least sine waves. The function generator might also produce square waves, sawtooth waves, triangle waves, or pulses, depending on the nature of the signal generator. When the CW mode is desired the modulating source is turned off.

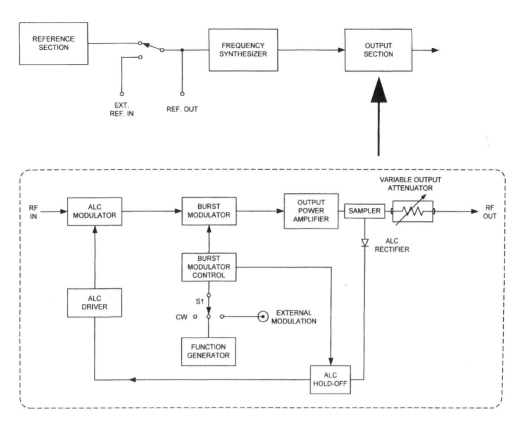

Figure 4.7 Detail of output section.

Note that there are two modulators in Figure 4.7: an *ALC modulator* and a *burst modulator.* The ALC modulator is designed to keep the output RF level constant, while the burst modulator wants to vary the output RF level in accordance with the applied audio signal. These two circuits are, therefore, in conflict with each other. The *ALC Hold-Off* circuit is used to moderate this conflict.

Frequency and Phase Modulation (FM and PM)

There are two basic forms of angular modulation: *frequency modulation* (FM) and *phase modulation* (PM). In FM the carrier frequency is varied in accordance with the modulating frequency, while in PM the carrier frequency remains constant but its phase is varied. Both FM and PM are considered essentially the same. Consider the case of FM. When there is no modulation present, the RF carrier (f_c) will remain constant. But when at a sine wave, audio signal is applied, the carrier frequency will deviate from f_c to a lower frequency (F1) on one peak of the modulating frequency (f_m), and to a higher frequency (F2) on the alternate peak of f_m. The *deviation* is the difference between the carrier and either extreme, that is, $+F_d = F2 - f_c$ and $-F_d = f_c - F1$. In most cases, signal generators use symmetrical sine wave modulation, so $+F_d = -F_d$, and is simply called F_d.

The peak voltage of the carrier remains constant and is defined by:

$$V = V(t)\ Sin[2\ \pi\ f_c\ +\ \beta\ m(t)] \qquad\qquad [4.6]$$

In FM the depth of modulation is expressed as the *modulation index* (β), which is defined as the ratio of the deviation to the modulating frequency, or F_d/f_m. The FM process produces a large number of sidebands, and at certain values of β the carrier will go to zero. The sidebands are described by mathematical entities called *Bessell functions.* In phase modulation the modulation index (β) is related to the variation in the phase, that is, $\beta = \Delta\varphi_{peak}$.

Straight frequency modulation is created by varying the frequency of the oscillator in the frequency synthesizer (see Figure 4.8). Phase modulation, on the other hand, is sometimes generated by using a *reactance modulator* to vary the phase of the oscillator signal.

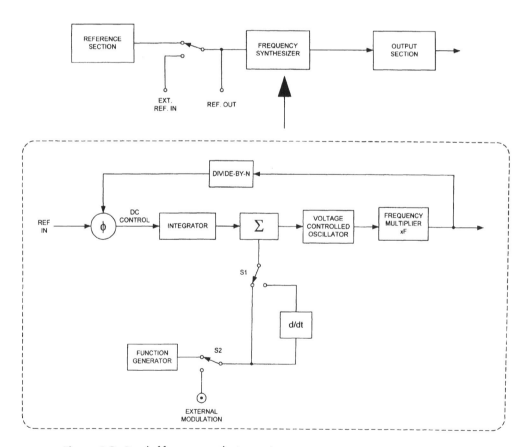

Figure 4.8 Detail of frequency synthesizer section.

SWEEP GENERATORS

Sweep generators ("sweepers") are used to produce a signal that changes frequency over a specified range. Although the sweeper nominally resembles an FM generator, there are differences. For one thing, the sweep range is usually quite a bit larger than the deviation of an FM signal generator. Another thing is that the sweeper tends to change frequency from one limit to the other, and then snaps back to the first limit.

There are three basic forms of sweep: *linear* (or "ramp") *sweep, stepped sweep,* and *list sweep.* All of these forms use a voltage-controlled oscillator frequency synthesizer, but the difference is in the waveform used for sweeping the output signal.

Figure 4.9A shows linear sawtooth waveform sweep. At time T1, the ramp applied to the VCO is at zero and begins ramping up. The frequency begins to move upwards from the zero-volts value in a linear manner until time T2, at which point it snaps back to the zero-volts value.

Stepped sweep (Figure 4.9B) uses a series of discrete voltage steps to change the frequency of the VCO. This method does not produce an output on every frequency in the output voltage range from T1 to T2, but rather only at specific values determined by the steps. The steps are produced in a circuit such as shown in Figure 4.10.

A digital-to-analog converter (DAC) produces an output voltage that is proportional to the binary number applied to its inputs. The maximum output voltage is set by an internal or external DC reference voltage, and the applied binary number. In this case, the DAC input is driven by a binary counter, which in turn is incremented by a digital clock (square wave generator).

The maximum number of steps that can be accommodated using any given binary counter is 2^N, where N is the bit length of the counter. For the 16-bit counter shown in Figure 4.10, therefore, a total of $2^{16} = 65,536$ different levels (including zero) can be created. The maximum output voltage is less than the reference voltage by the 1-LSB (least significant bit) voltage, or $1/2^N$. If a 16-bit counter is connected to a DAC with a 10.00 volt reference source, then the 1-LSB step is 0.00015 volts and the maximum output voltage is $10.00 - 0.00015$ volts $= 9.99985$ volts.

Output Reverse Power Protection

Many signal generators are used to test radio communications transceivers. This equipment contains both transmitter and receiver, which share a common output connector to the antenna. If the signal generator is connected to the antenna connector to test the receiver, and someone accidentally keys the transmitter, then serious damage can result to the signal generator. High-quality instruments provide architectures with *output reverse power protection.* This circuitry will prevent damage to the signal source if a high RF power level is applied to its output connector. List sweepers also have a DAC to control the VCO, but will drive it with a series of values stored in a digital memory similar to those used in computers. This permits arbitrary waveforms to be created.

Figure 4.11 shows one way that sweepers are built: *open loop.* The frequency synthesizer loop is broken, and a linear sawtooth, stepped waveform, or

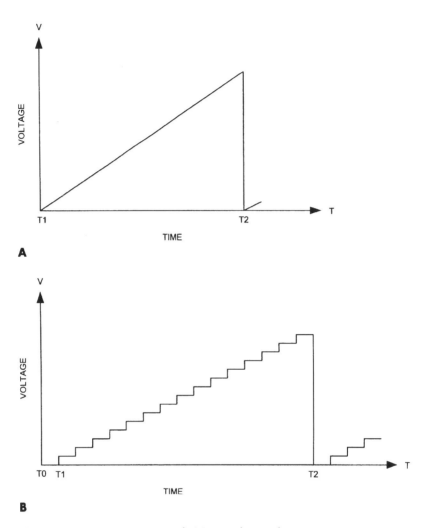

Figure 4.9 (A) Linear ramp sawtooth; (B) Stepped sawtooth.

list-generated waveform is applied to its DC control voltage input. Another approach is shown in Figure 4.12. This is a *closed loop* approach. The frequency synthesizer loop is not opened, but rather a binary counter or list memory is used to drive the divide-by-N counter in the synthesizer PLL feedback loop.

IMPROVING THE QUALITY OF SIGNAL GENERATOR USE

Improving Spectral Purity

Certain high-quality measurements are very sensitive to extraneous signals coming out of a signal generator. Many signal generators put out harmonics that are

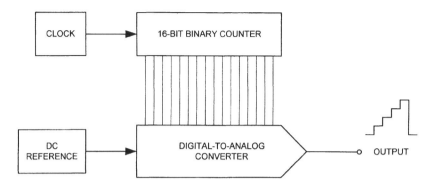

Figure 4.10 Generating a stepped sawtooth using a binary counter and digital-to-analog converter.

−30 dB down from the main signal ("carrier"), while other signals may be either higher or lower than this level. The way to get rid of these extraneous signals is to place a frequency-selective filter between the output of the signal generator and the device under test (DUT).

Figure 4.13 shows the use of a low-pass filter to eliminate the harmonics and any spurs that are above the main signal. Select a filter with a −3 dB point somewhere between the main signal and the first extra signal, and an attenuation slope large enough to reduce the "bad" signals as much as possible. If there are any subharmonics (or spurs lower than the main signal), then either use a bandpass filter or add a high-pass filter with a −3 dB cut-off between the main signal and the subharmonic.

There is a cautionary note, however. Real filters do not have the nice flat response seen in some textbooks. They will have passband ripple, and some odd responses out of band. Also, L-C filters are notoriously unpredictable when you terminate them in an impedance other than the design impedance. Understand the passband response and the insertion loss of the filter before using it.

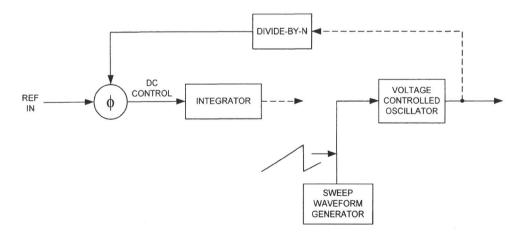

Figure 4.11 Analog sweeper block diagram.

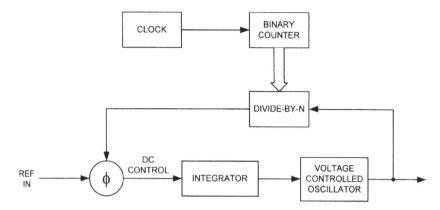

Figure 4.12 Digital sweeper block diagram.

Improving Mismatch Loss

Mismatch error occurs because the load and the signal generator are not impedance matched. In any electronic circuit, the maximum power transfer occurs when
the impedances are matched. There may be an inherent mismatch problem in either the signal generator or the load, and almost certainly in the cables or other devices connected in line with the signal generator.

For example, assume we have a signal generator with a VSWR of 1.9:1, and
a DUT (Device Under Test) with a VSWR of 1.6:1 connected in the normal
way (Figure 4.14A). The mismatch loss can be found once we know the reflection
coefficients:

Source:

$$\rho_s = \frac{SWR1 - 1}{SWR1 + 1} = \frac{1.9 - 1}{1.9 + 1} = \frac{0.9}{2.5} = 0.31$$

DUT:

$$\rho_D = \frac{SWR1 - 1}{SWR1 + 1} = \frac{1.6 - 1}{1.6 + 1} = \frac{0.6}{2.6} = 0.23$$

Mismatch Loss:

$$\text{M.L.} = 20 \, \text{LOG} \left[1 + (\rho_S \rho_D) \right]$$

$$= 20 \, \text{LOG} \left[1 + ((0.31)(0.23) \right] = 20 \, \text{LOG} \left[1.07 \right]$$

$$= (20)(0.03) = 0.6 \, \text{dB}$$

Figure 4.14B shows the way of dealing with this problem. Insert a 10-dB fixed attenuator in line with the line between the signal generator output and the DUT. You
will have to adjust the signal-generator output-level control 10-dB higher than normal to compensate for the extra attenuation. The reason this works is that fixed resistive attenuators tend to be designed with very low reflection coefficients.

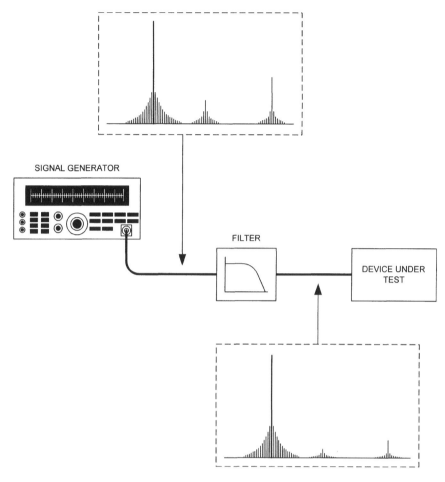

Figure 4.13 Improving output spectrum using a filter.

Suppose we have the same components in Figure 4.14B as in Figure 4.14A, but add an attenuator with $\rho_A = 0.31$. The mismatch loss becomes:

$$M.L. = 20 \ LOG \ [1 + (\rho_S \ \rho_D \ (\rho_A)^2)]$$

$$= 20 \ LOG \ [1 + ((0.31)(0.23)(0.31)^2)]$$

$$= 20 \ LOG \ [\ 1 + 0.0069] = 20 \ LOG \ [1.0069]$$

$$= (20)(0.003) = 0.06 \ dB$$

Improving Third-Order Intercept Performance

One of the most important specifications for an amplifier or radio receiver is the *third-order intercept point* (TOIP or IP3). This specification tells us something about the

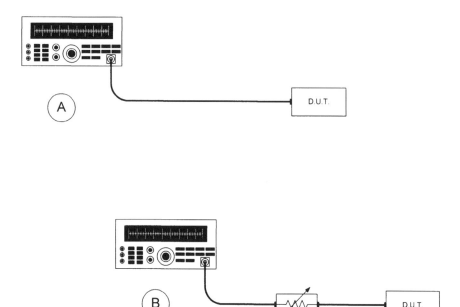

Figure 4.14 (A) Connecting the generator directly to Device Under Test (DUT); (B) Using a variable or switched attenuator in signal line between generator and DUT.

device's dynamic performance, especially in the presence of multiple input signals. If you listen to any shortwave receiver, an AM or FM BCB receiver, or any scanner receiver, you will realize that most areas of the country are polluted with too many radio signals. When multiple strong signals are received at the same time, receiver (or amplifier) nonlinearity will occur, and heterodyne products are created. If F1 and F2 are two input signals (one of which might be the desired signal), these "intermodulation" products will have frequencies equal to mF1 ± nF2, where m and n are integers. The third-order harmonics are those in which m = 2 and n = 1, or m = 1 and n = 2 (i.e., m + n = 3), and these are the most difficult to handle. The worst case is usually the 2F1 − F2 and 2F2 − F1 third-order products, because they will fall close to F1 and F2 and may be within the device passband. A problem with these products is that they increase at a rate three times the number of decibels as the fundamental signal. If F1 or F2 goes up 1 dB, then the third-order products go up 3 dB.

Figure 4.15 shows the basic set-up for measuring the third-order intercept point (also certain other parameters). The two-signal generators produce F1 and F2. They are set to identical output levels, which are usually quite high, such as −10 dBm or −20 dBm. Initially, the receiver is tuned to one frequency (e.g., F1), and a reference level is established that is equal to the minimum discernible signal (MDS), or in some procedures an S1 signal level. The receiver is then tuned to the third-order product frequencies, and the attenuator decreased (raising signal level) until the same reference level is produced. The IP3 can be calculated from these data points.

But look what happens if the signal generator is working quite the way we hoped (Figure 4.16). The spectrum of Figure 4.16A is what we hope to see. Frequencies F1 and F2 are standing up smartly above the noise level (which is hopefully quite low). Figure 4.16B shows the same spectrum with the third-order IM

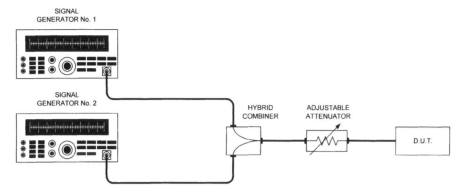

Figure 4.15 Intermodulation test set-up (two-tone).

products present. When you see this at the output of an amplifier or receiver being tested, then you might assume that the IM products are generated in the DUT. But this is not always the case. Sometimes, the signal from one generator gets into the output stages of the other generator and causes an IM response that is due solely to the test set-up!

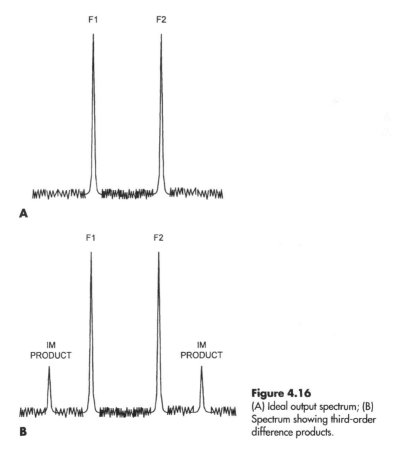

Figure 4.16
(A) Ideal output spectrum; (B) Spectrum showing third-order difference products.

A couple things can be done to prevent the problem. First, if the combiner used to merge the signals into one line is a resistive star type circuit, then there is only 6 dB of isolation between the ports. Using a hybrid combiner with a larger amount of port-to-port isolation helps tremendously, because it reduces the signal reaching the other generator's output stages.

Another fix is to insert 10-dB or 20-dB fixed attenuators in each signal line (Figure 4.17). These attenuators provide an additional amount of isolation between the signal generators. Of course, you will have to adjust the output levels of the signal generators to overcome the extra loss.

A cautionary note is in order: Be certain that the signal generator output can be cranked up to a higher level without producing spurious output signals, harmonics, and other extraneous signals. One of my own signal generators works well from 0 to 90% of full output, but at output levels > 90%, the spectrum blossoms with unwanted signals.

Extending Upper Output Range

Signal generator outputs controls are calibrated in terms of output voltage, usually microvolts (μV) or millivolts (mV), or the power level (e.g., dBm—decibels relative to 1 mW in 50 ohms). A typical generator produces output levels up to some value from about 0 dBm, or perhaps + 20 dBm (or some value in between). But what do you do if the signal generator maximum is, say, + 10 dBm, and you need a signal level of + 30 dBm (1 watt)? Or, how about the case where you have a signal generator like mine that is wonderful at lower levels, but falls apart at higher levels?

The solution is simple and obvious: amplify the output. But there are some cautions. Figure 4.18 shows the use of an external amplifier to boost the output level of the signal generator. Because all amplifiers can become nonlinear, and produce a bit of harmonic distortion in their own right, a low-pass filter is inserted in the path between the amplifier output and the DUT.

You must also make sure that the selected amplifier can do the job. Make sure the IP3 specification of the amplifier is sufficiently high that the signal generator cannot overdrive it. The maximum input drive level (usually specified in dBm) and

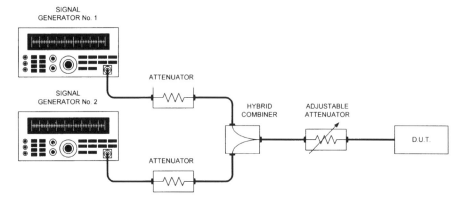

Figure 4.17   Improved IMD testing scheme.

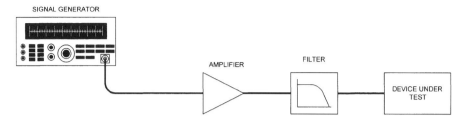

Figure 4.18 Increasing output power of the signal generator using external amplifier. Low-pass filter reduces noise and spurious products produced by the amplifier.

the output power level (also expressed in dBm or possibly watts) must be sufficient to handle the job. Otherwise, adding the amplifier might add problems as well.

Reducing the Output Level

There are times when you might want to reduce the signal generator output level. One reason for doing this is that you need a very small signal at the DUT, but you need a higher signal to act as a reference or be fed to a frequency counter.

Sometimes this is done when using an elderly analog signal generator that has an inaccurate analog frequency dial, but the output attenuator is well calibrated. To get a higher level for the counter, while providing a low-level signal to the amplifier or receiver being tested, use a set-up like the one in Figure 4.19. In other cases, you will simply need a lower signal level than the generator can provide. Figure 4.18 also works for that purpose as well.

The attenuator should be a calibrated type. You can obtain continuously variable calibrated attenuators, but these are costly (although some tend to come on the surplus market). A lower-cost alternative is to use a precision step attenuator. These devices have switch-selectable attenuation levels in various steps. The total attenuation is the sum of all the individual attenuations. You can build step attenuators, but for precision work you are well advised to buy one (the resistors for precision attenuation levels are really odd values, although they can be approximated).

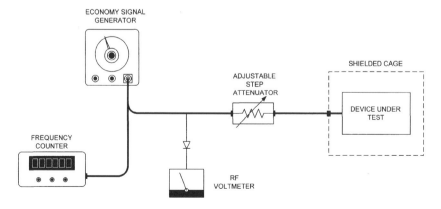

Figure 4.19 Using an economy or service grade signal generator to measure a sensitive device or circuit.

CHAPTER FIVE

Spectrum and Network Analyzers

Signals can be represented in a number of ways. The most familiar is the time-domain representation shown in Figure 5.1A. This view of a pair of signals plots their amplitudes against time, thus revealing their waveshapes (in this case sinusoidal). From an amplitude-versus-time display one can tell the frequency (because $F = 1/T$), amplitude, and waveshape. An oscilloscope is normally used to view the time domain aspect of a signal.

Another view is the frequency domain shown in Figure 5.1B. This display plots amplitude-versus-frequency, so the same two signals seen in Figure 5.1A will plot as a pair of spikes in Figure 5.1B. The comprehensive view of signals requires that we take a look at both time domain and frequency domains. Because they share a common axis and amplitude, we can view them orthogonally as in Figure 5.1C.

All continuous waveforms can be represented mathematically by a series of sine and cosine functions. Only the sine wave is pure in the sense that it contains only one frequency. All other waveforms, including sine waves with even the smallest possible amount of distortion, possess a number of harmonically related frequency components. The specific harmonics, their amplitudes and phases, determine the final shape of the overall complex wave. The complex wave can be described by a *Fourier* series of the form:

$$f(t) = \frac{a_o}{2} + \int_{n=1}^{\infty} (a_n \, Cos(n\omega t)) + (b_n \, Sin(n\omega t))$$

Where:

    a_n and b_n represent the amplitudes of the harmonics (see below)
    n is an integer
    f is the frequency in Hertz
    ω is the angular frequency $2\pi f$

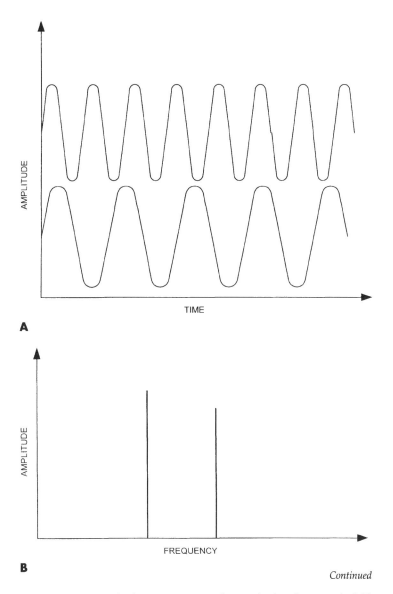

TIME

A

FREQUENCY

B

Continued

Figure 5.1   (A) Amplitude-versus-time (time-domain) display of two signals of different frequencies as seen on an oscilloscope; (B) Amplitude-versus-frequency (frequency-domain) display as seen on a spectrum analyzer; (C) Combined time-domain and frequency-domain displays.

The amplitude coefficients (a_n and b_n) are expressed by:

$$a_n = \frac{2}{T} \int_0^t f(t)\, \cos(n\omega t)\, dt$$

Figure 5.1 (*continued*)

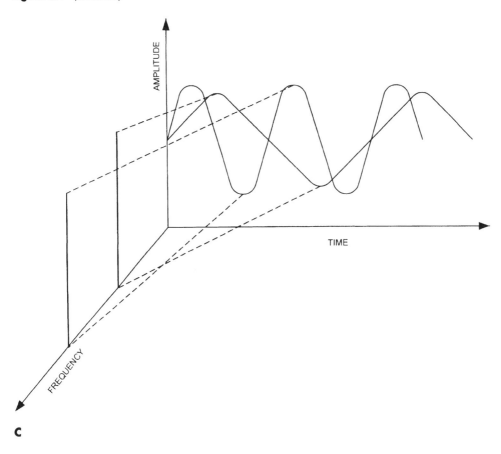

c

and,

$$b_n = \frac{2}{T} \int_0^t f(t) \sin{(n\omega t)}dt$$

The amplitude terms are non-zero at the specific frequencies determined by the Fourier series. Because only certain frequencies, determined by integer n, are allowable, the spectrum of the periodic signal is said to be *discrete*.

The term $a_0/2$ in the Fourier series expression is the average value of f (t) over one complete cycle (one period) of the waveform. In practical terms, it is also the *dc component* of the waveform.

USING THE FREQUENCY DOMAIN

If we were certain that all signals in a system were pure sine waves, that there was no modulation or heterodyne mixing taking place, and that all stages in the system were perfectly linear, then the time-domain display seen on an ordinary oscillo-scope would suffice for practical purposes. But that never happens. Real signals have distortion, undergo both modulation and frequency mixing, and never see a perfectly linear signal-processing stage.

The principal uses of a spectrum analyzer are to examine *noise, distortion, mixing action,* and *modulation.* It is necessary to characterize signals going into and coming out of a system in order to understand how the system acts on the signal. By examining the "goes-intos" and "goes-out-ofs" we can characterize the system and determine its performance.

Noise

Figure 5.2A shows a frequency-domain characterization of a noise signal. Understanding the noise spectrum allows us to either evaluate or design the system to best overcome its effects. The noise spectrum permits us, therefore, to spot problems in system performance and design accordingly.

Harmonic Distortion

When a pure sine wave is passed through a nonlinear stage harmonic components are generated. These new frequencies are integer multiples of the fundamental frequency (2F, 3F, 4F . . . nF). When a non-pure sine wave (which has its own harmonics) is processed in a nonlinear stage, additional harmonics or increased harmonic amplitudes are created. Figure 5.2B shows the frequency spectrum of a waveform with multiple harmonics present. The tallest spike represents the fundamental frequency sine wave, while the smaller spikes are the harmonics.

Intermodulation Distortion

While harmonic distortion occurs on a single signal, intermodulation distortion (IMD) occurs when two or more signals mix in a nonlinear circuit. When this occurs, additional frequencies are generated according to the rule $mF1 \pm nF2$, where m and n are either zero or integers. Figure 5.2C shows this action when two equal amplitude signals (F1 and F2) interact in a nonlinear manner. The two small peaks are particularly interesting in amplifier and receiver designs because they fall close to F1 and F2 (other products fall very far away). These are the 2F1-F2 and 2F2-F1 products.

When undesired, this effect is called IMD, but when it is desired to translate frequencies in a mixer circuit, the effect is called heterodyning.

Modulation

A single-frequency unmodulated signal will have a spectrum consisting of a single spike in the absence of distortion. But when information is imparted to the signal, additional products are created. These show up as a spectrum similar to Figure 5.2D. Here we see the result when a sine wave RF carrier is amplitude-modulated by a sine wave audio tone. In this case, a mixing action takes place, as shown clearly on the spectrum trace. If an RF carrier F_c is modulated by an audio tone F_a, then the two products are $F_c + F_a$ and $F_c - F_a$. These are called the upper and lower sidebands (USB and LSB), respectively.

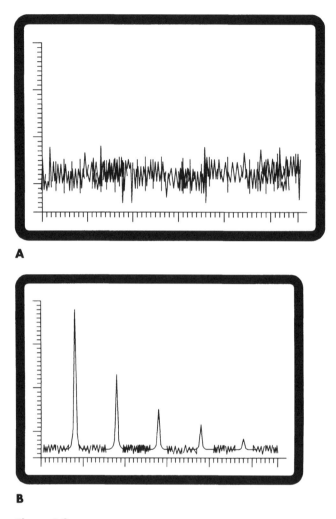

Figure 5.2
(A) Signal buried in noise; (B) Harmonic distortion; (C) IM distortion;
(D) AM signal spectra.

SPECTRUM ANALYSIS

Spectrum analysis becomes possible when the various frequency components and noise are measured and displayed. Over the years several approaches have been taken to spectrum analysis: Fourier analyzer, tunable filters, and spectrum analyzers.

The Fourier analyzer is depicted in Figure 5.3A, while its display is shown in Figure 5.3B. The analyzer consists of a series of adjacent bandpass filters, each of which passes a small amount of spectrum. When the outputs of these filters are poled, it is possible to build the display shown.

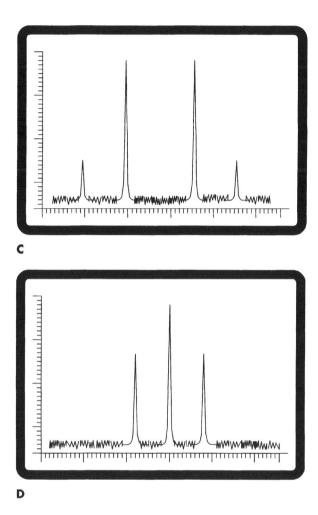

C

D

There are a number of problems with the Fourier analyzer. First, it's not terribly flexible because the filters are fixed tuned. Second, the resolution depends on the filter bandwidth, which may not be consistent throughout the range of frequencies being measured. Finally, there are a restricted number of adjacent frequency filters that can be accommodated, especially where cost is a consideration. Finally, there may be interaction between the filters, causing a loss of performance.

The tunable filter approach is shown in Figure 5.4. A filter is designed to be tuned across the entire range of frequencies by manual means. In most cases, these instruments were actually special-purpose radio receivers. When the output was calibrated, the instrument was called either a wave analyzer or tunable RF voltmeter.

Both of these approaches suffer from major faults, not the least of which is poor ease of operation. The modern spectrum analyzer solves these problems rather nicely.

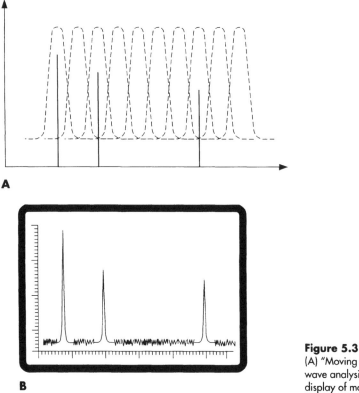

Figure 5.3
(A) "Moving filter" approach to wave analysis; (B) Oscilloscope display of moving filter results.

THE SPECTRUM ANALYZER

The spectrum analyzer basically automates and improves the tunable RF voltmeter. Figure 5.5 shows the basic block diagram for a generic spectrum analyzer. It is a narrow band receiver that is swept-tuned across the range of interest. A sawtooth ramp waveform is used to sweep-tune the receiver and to drive the horizontal deflection system of the oscilloscope. The output of the receiver is proportional to signal strength; it is also applied to the vertical input of the oscilloscope. The result is the spectrum plot shown.

To understand the operation of the spectrum analyzer let's take a look at each stage in its turn. The heart of the spectrum analyzer is the *mixer* and *local oscillator* (LO).

The LO is a voltage-controlled oscillator (VCO) that produces an output frequency that is proportional to an applied input control voltage. In the case of the spectrum analyzer the input voltage is a ramp, so the voltage will change as the ramp voltage rises. Because most VCO circuits have a quadratic relationship between the control voltage and frequency, it may be necessary in some cases to alter the tuning voltage waveform from a linear ramp to a shape that makes the sweep of the VCO output frequency look linear.

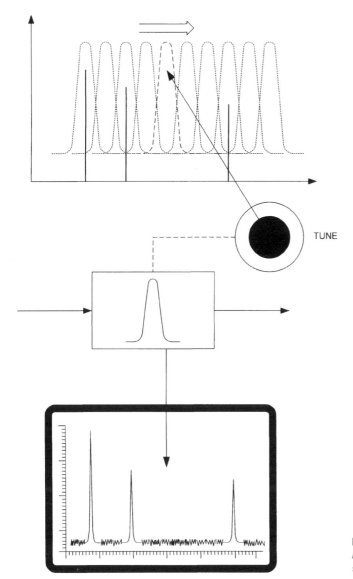

Figure 5.4
Manual wave analysis preceded spectrum analyzers.

The mixer is a nonlinear circuit that mixes the RF input signal (F1) with the LO signal (F2) to produce an intermediate frequency (IF) output. Any of the frequencies described by $mF1 \pm nF2$ can be used, but it is not reasonable to use other than the second-order products (F1 + F2 or F1 − F2).

The characteristics of the mixer are important to the quality of the spectrum analyzer. Double-balanced mixers are usually preferred over single-ended or single-balanced mixers because they tend to cancel the F1 and F2 signals in the output, leaving only the sum and difference products. Other forms of mixers invariably have a residual F1 and/or F2 component present in the IF output port.

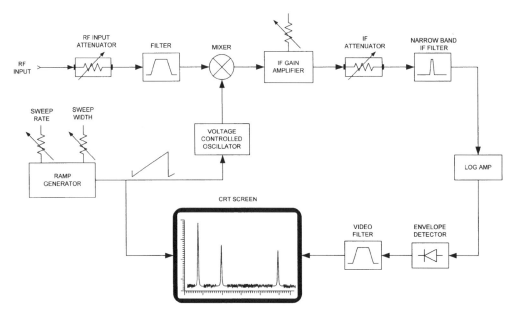

Figure 5.5 Block diagram of spectrum analyzer.

A properly designed mixer circuit should be terminated in its characteristic impedance (e.g., 50 ohms). Because only one of the two second-order products is needed, mixer performance is improved if the unwanted second-order product is absorbed, rather than reflected back towards the mixer IF port. A circuit called a *diplexer* is often used to separate the two second-order products, and to route the desired product to the IF amplifier and the undesired product to a matching dummy-load resistor.

It is very important to select a mixer with a high third-order intercept point (also known as IP3 or TOIP) and a high dynamic range. One of the failings of cheap spectrum analyzers is that the mixer does not possess these attributes, so it is possible to generate both harmonic and intermodulation distortion products inside the mixer. They will appear at the output of the spectrum analyzer, and be displayed on the screen, despite the fact that they are spurious signals not present in the input spectrum.

The front-end of the spectrum analyzer consists of the mixer/LO plus any preprocessing that must be done. There are two forms of preprocessing shown here: *RF input attenuator* and *RF filter*. Some spectrum analyzers might also have a preamplifier. It is not unreasonable to expect these stages to be switch selectable.

The RF attenuator is used to reduce the amplitude of all signals applied to the RF input of the spectrum analyzer by an equal amount. The input attenuator is used to prevent the mixer and any preamplifiers used from going into gain compression. Once gain compression is reached, IM products begin to creep upwards, distorting the picture of the spectrum with spurs that were not in the original.

The input filter may be a bandpass, low-pass, or high-pass filter, and is used to prevent unwanted frequencies from entering the spectrum analyzer. If you are

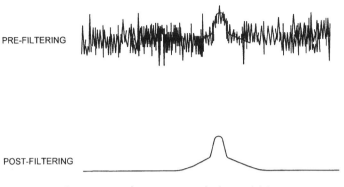

PRE-FILTERING

POST-FILTERING

Figure 5.6   Filtering improves display readability.

looking at a fairly limited range of frequencies, such as the modulation around a transmitter signal, then filtering can eliminate out-of-range signals from interfering with the process. Those unwanted signals could conceivably force the mixer into gain compression and create spurs.

The IF section handles the signal from the output of the mixer. Most of the gain and selectivity of the spectrum analyzer is provided in the IF section. The principal stages are: *IF gain amplifier, IF attenuator, narrow-band filter,* and *logarithmic amplifier.*

The gain amplifier provides adjustable voltage gain to permit adjustment and scaling. It is used to adjust the vertical displacement of the signals without changing the input conditions. In some cases, the amplifier gain and input attenuator are adjusted in tandem to prevent shifts in the vertical display.

The *resolution bandwidth* (RBW) of the spectrum analyzer is set by the IF filter. The smallest resolvable frequency unit is determined by the RBW, so making it too wide will cause smearing of close-in responses.

The logarithmic amplifier serves to provide range compression so that both high- and low-amplitude products can be displayed at the same time. Otherwise, providing enough gain to see low-amplitude signals will cause high-amplitude signals to go off the top of the scale. Also, using a logarithmic amplifier permits the use of decibel notation on the CRT screen.

The *video detector* is an envelope detector that produces a DC output that is proportional to the signal strength at the frequency being measured. It produces the vertical deflection signal seen on the CRT screen. But this signal is often not too clean and must be filtered. Figure 5.6 shows pre-filtering and post-filtering versions of a noisy waveform. The pre-filtering version is barely usable, if that.

RESOLUTION

Resolution is the ability to distinguish two things, so in spectrum analyzers we can define resolution as the measure of the instrument's ability to distinguish between two adjacent frequencies. This concept is shown in Figure 5.7. Although Figure 5.7 exaggerates the issue for the sake of illustration, the concept is the same. In Figure 5.7A the bandwidth of the system is set too broadly. Adjacent frequencies F1 and F2 are not discriminated, but appear as one smeared trace on the CRT screen. Narrowing the bandwidth a little produces a pattern such as Figure 5.7B. A dip appears in the pattern, indicating that there might be two frequencies present, but imparting little additional information. In Figure 5.7C the bandwidth is narrow enough to break out the two frequency spikes.

There are several things that affect the resolution of the spectrum analyzer. One is the *selectivity* (Figure 5.8). Selectivity is the ability to resolve adjacent signals. Resolution is usually stated in terms of breaking out two adjacent equal amplitude signals, while selectivity addresses the ability to break out two adjacent unequal signals. In the latter case a small signal nestled close to a larger signal could be lost in the filter skirts of the larger signal. Figure 5.8 shows the definition of selectivity: it is the ratio of the −60 dB bandwidth to the −3 dB bandwidth, or

$$Selectivity = \frac{-60\,dB\ B.W.}{-3\,dB\ B.W.} \qquad [5.1]$$

Typically, the value of selectivity will be around 5:1 for digital filters, and between 10:1 and 15:1 for analog filters. Consider an example. Two signals, F1 and F2, are spaced 20 kHz apart. Suppose the resolution bandwidth (RBW) is 3 kHz, and the filter has a selectivity of 14:1. The bandwidth at −60 dB is, therefore, (−3 dB BW) × Selectivity = 3 kHz × 14 = 42 kHz. Because the selectivity is much larger than the frequency spacing, one signal is lost in the skirts of the other. If there is a large amplitude difference (e.g., 60 dB), then the lower signal will be lost in the noise. A rule of thumb is that *minimum discernible frequency separation for highly unequal signals is one-half the −60 dB bandwidth.*

The bandwidth of the system is a key player, but it is not necessarily solely the IF bandwidth. For example, it is quite reasonable to make the video resolution bandwidth one-tenth the IF resolution bandwidth in order to improve discrimination. Other factors also affect resolution and display. For example, sweep-speed/measurement time, residual FM and phase noise, and noise floor.

The RBW of the spectrum analyzer is set by internal filters, most profoundly in the IF amplifier and video amplifier sections. These filters have an associated time constant required for them to correctly respond to an input signal. If the spectrum analyzer LO sweeps by too fast for the set RBW then we see smearing of the signal, amplitude error, and error in the frequency of the signal being measured. It is also possible for a too-fast sweep to cause "ringing" in the filters. The narrower the filter's bandwidth, and the better the selectivity curve of the filter, the more likely that fast-sweeped LOs will cause ringing, and therefore distortion of the signal being displayed. On the other hand, too slow a sweep will cause uncomfortable flicker in the display, making it more difficult to either interpret or photograph. Sweep speed, therefore, is a trade-off between the needs of the filter and the needs of the display.

Residual FM in the spectrum-analyzer-swept LO tends to smear the signals displayed and is not usual detectable by the user. The key is to specify very low

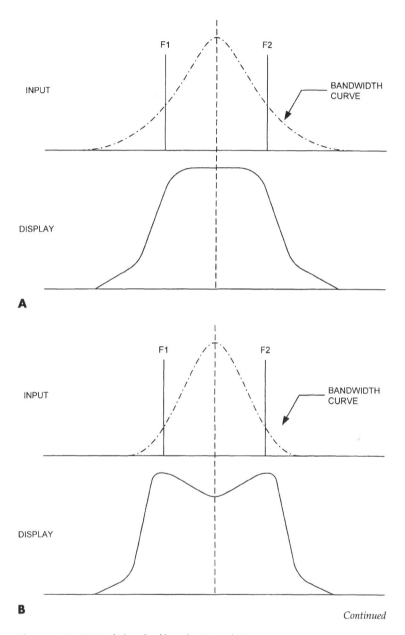

Figure 5.7 (A) Wide bandwidth makes F1 and F2 appear as "on target"; (B) Narrowing bandwidth makes peaks begin to appear, but they are still obscured; (C) Sufficiently narrow bandwidth allows the two signals to be separated.

levels of residual FM, which is an argument in favor of using a phase-locked loop-frequency synthesizer (see Chapter 3, "Smith Charting the Radio Frequency Circuit," for a description of these circuits).

Another factor is the *noise sidebands* or *phase noise* of the system. Figure 5.9 shows the phase-noise problem. All signal sources produce some amount of

Figure 5.7 (continued)

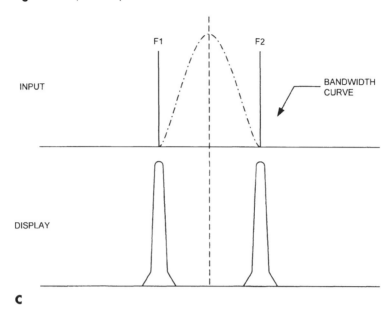

C

unintended phase noise. The level of noise is usually measured in decibels below carrier (dBc), and is normalized to a 1-Hz bandwidth (i.e., dBc/Hz). For
example, suppose we need to measure a distortion product signal that is −60 dB
below a carrier using a 3-kHz RBW. To normalize the phase noise in 3-kHz in
terms of dB:

$$Phase\ Noise = 10\ LOG_{10}\left[\frac{BW\ (Hz)}{1 - Hz}\right] \qquad [5.2]$$

So, for 3-kHz bandwidth $10 \times LOG_{10}[3000/1] = (10)(3.48) = 34.8$ dB. Thus, to view
a signal that is −60 dB down from a carrier at the input, the spectrum analyzer will
need a phase noise specification of −60 dBc − (34.8 dB) = −94.8 dBc.

All electronic instruments produce noise, and spectrum analyzers are no ex-
ception. If you either short circuit the RF input, or terminate it in a matched im
pedance, then you will still see a noise level present on the CRT display. In the case
of a properly terminated input (e.g., with a shielded 50-ohms resistor in place) the
noise level will be higher than the expected thermal noise (4 KTBR). The excess
noise is created inside the instrument.

A practical aspect of the internal noise generated is that it is not affected by
the RF input attenuator, while the signal is so affected. Figure 5.10 shows the effect
on a weak-signal display (i.e., the signal is close to the noise level). In Figure 5.10A
the signal sees 0 dB attenuation in the RF front-end and is discernible above the

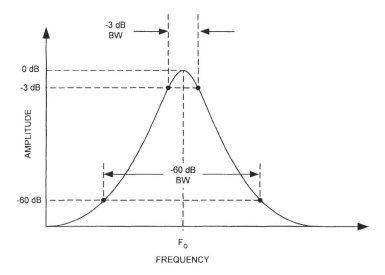

Figure 5.8 Bandwidth definition is between −3 dB points, while the shape factor is the ratio of the −3 dB and −60 dB bandwidth.

noise floor. But in Figure 5.10B the RF attenuator is set to reduce the input signal −10 dB, but the noise level remains the same. The signal is now obscured by the noise floor of the spectrum analyzer. The noise floor is specified in terms of *displayed average noise level* (DANL) and is a function of bandwidth (see Chapter 8, "Radio Receivers and Their Measurements," for further discussion of the relationship between noise floor and bandwidth).

Because the noise floor is a function of bandwidth, the situation can often be improved in a practical sense by narrowing the bandwidth to the minimum necessary to correctly show the signal (excess bandwidth increases DANL). In addition, it is also wise to keep the sensitivity low enough that the signal is readable and measurable, but so the noise floor is reduced to a negligible level.

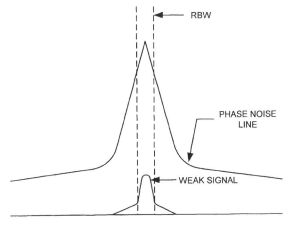

Figure 5.9
Effect of real bandwidth on displayed signal.

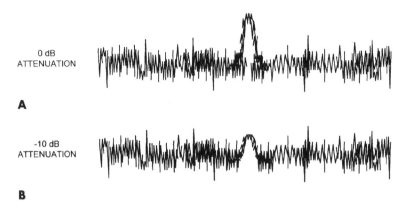

Figure 5.10 Adding attenuation inappropriately can cause signal to disappear into the noise.

USING THE SPECTRUM ANALYZER

There are several different ways to use a spectrum analyzer. Figure 5.11 shows the use of the spectrum analyzer to view the output of a signal source or signal generator. This will work, but is not necessarily the best way to approach the problem. For one thing, if the output level of the signal generator is too high it might drive the spectrum analyzer into distortion. The displayed spectrum would then exhibit elements that are due to the test set-up, not to what is actually coming from the signal generator. The problem can be partially overcome either by keeping the signal level low enough, or placing attenuators in the line between the signal source and spectrum analyzer.

Another application is to measure the output spectrum of some circuit, or "device under test" (DUT). This approach is used to measure both the harmonic and intermodulation distortion produced by the DUT. Unfortunately, the spectrum analyzer may produce some distortion of its own, and because the same fundamental signals are involved, it will look like distortion produced in the DUT. A simple test to show whether an observed distortion product is due to the spectrum analyzer or the DUT is to change the attenuation applied to the RF front-end and

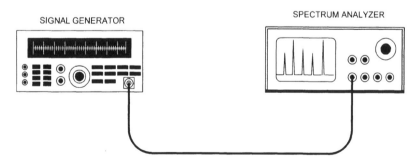

Figure 5.11 Measuring output spectrum of signal generator.

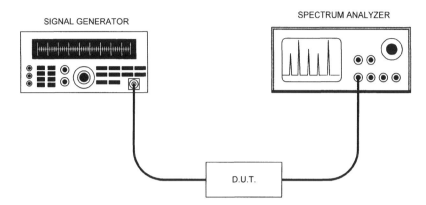

Figure 5.12 Measuring device under test spectrum output. Compare the spectrum seen in this test with that of Figure 5.11 in order to ferret out signal generator contribution.

observe the effects. If the input attenuation is changed by 10 dB, and the displayed distortion products change, then at least part of the distortion is being created in the spectrum analyzer. If the level of the products does not change, then the products are due to the DUT, not the spectrum analyzer.

Small transmitters can be tested off-the-air by keying them in close proximity to a spectrum analyzer that is fitted with a small whip antenna (Figure 5.13). Care must be taken to keep the transmitter close enough to produce a readable display, but far enough away not to overload the spectrum analyzer. It is possible for higher-power transmitters, or transmitters brought too close to the spectrum analyzer antenna, to damage the analyzer input. If the transmitter is too far from the spectrum analyzer antenna, then other signals may be visible on the display as well as the desired signal. That is a constant concern when doing off-the-air checks with a sensitive spectrum analyzer.

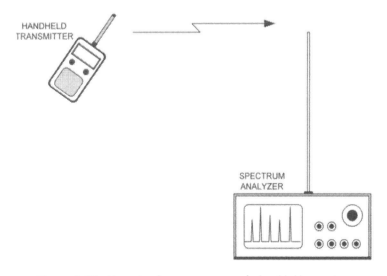

Figure 5.13 Measuring the output spectrum of a hand-held transmitter.

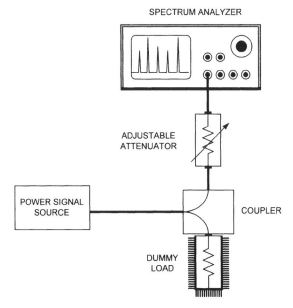

Figure 5.14
Measuring spectrum output of a fixed transmitter may require a dummy load to absorb power from the transmitter, an isolating coupler to sample the signal, and additional attenuation to protect the spectrum analyzer.

Devices that produce RF output power levels that are too high for direct application to the spectrum analyzer can be tested as in Figure 5.14. In this arrangement the output power from the source is connected to a coupler with an isolated port as well as a straight-through or direct port. A dummy load is connected to the direct port, while the isolated port is fed to the spectrum analyzer. In some cases, an additional attenuator is needed to augment the isolation loss of the coupled port. Additional information on couplers is found in Chapter 2, " Small Components Used in Radio Frequency Test and Measurement."

Figure 5.15 shows a method used by some instruments to sample a high-power signal directly from a dummy load. Some dummy loads include the sampling loop ("gimmick"), while in others it has to be built. The small wire-sampling loop is used to pick up a tiny sample of the RF power applied to the dummy load resistor.

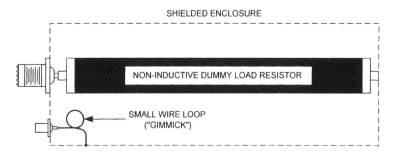

Figure 5.15   One method of providing a sample signal is by use of a "gimmick" loop inside a dummy load.

NETWORK ANALYZERS

A network analyzer examines incident, reflected, and transmitted signals through a circuit or device, and displays the magnitude and phase of these signals. A spectrum analyzer, on the other hand, measures only one channel, and displays magnitude and frequency. Figure 5.16 shows the optical analogy to the network analyzer. A "device" of different optical density than ambient is in the path of an incident ray of light (R). When the light hits the surface, part of it is reflected (A) and part is transmitted through the "device" (even though refracted a bit).

A network analyzer consists of a three-channel RF receiver and a display (Figure 5.17). The incident signal is considered the *reference signal,* so it is designated the R-channel. The other two channels receive the reflected signal on the A-channel and the transmitted signal on the B-channel.

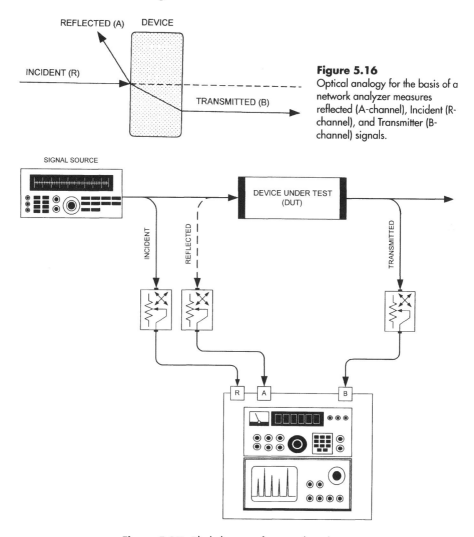

Figure 5.16
Optical analogy for the basis of a network analyzer measures reflected (A-channel), Incident (R-channel), and Transmitter (B-channel) signals.

Figure 5.17 Block diagram of a network analyzer.

The uses of scaler and vector network analyzers (Figure 5.18) differ from the uses of spectrum analyzers. The spectrum analyzer measures external signals of unknown frequency and modulation type. Even when a tracking generator is added, to allow the spectrum analyzer to perform stimulus-response tests, the spectrum analyzer cannot do the job of the network analyzer. The network analyzer, by contrast, contains a known signal source and is capable of sweeping a range of frequencies and power output levels. It can also perform ratio measurements.

Figure 5.18 Example of a vector network analyzer. (Photo courtesy of Hewlett-Packard)

CHAPTER SIX

Radio Frequency
Power Measurements

Radio frequency power measurements are made for a variety of purposes. In this chapter the several different topics will be discussed: the type of power being measured, methods of measuring power, error sources in RF power measurement, and typical commercial instruments used for RF power measurements.

The assumption in this chapter is that the RF power is being measured to determine the output level produced by some circuit or device. Chapter 11, "Antenna Gain and Pattern Measurements," will deal with RF power measurements made with respect to antennas. This chapter is, however, highly relevant to the material in Chapter 11.

WHAT IS POWER?

Electrical power is defined as *energy flow per unit of time.* The internationally accepted standard unit of power is the *watt* (W), which is defined as an energy flow of *one joule per second.* Other electrical units are defined in terms of the watt; for example, one volt is one watt per ampere of current flow. The watt is the product of the electrical potential and the current flowing:

$$P = V \times I \qquad [6.1]$$

Other expressions of power include:

$$P = I^2 \times R \qquad [6.2]$$

and,

$$P = \frac{V^2}{R}$$ [6.3]

Where:

P is the power in watts (W)
V is electrical potential in volts (V)
I is current in amperes (A)
R is resistance in ohms (Ω)

Subunits and Superunits

The unit *watt* is not always suitable for some applications. As a result, it is common to find subunits and superunits in use. These are:

Units	Size
Femtowatts (fW)	10^{-15} Watts
Picowatts (pW)	10^{-12} Watts
Nanowatts (nW)	10^{-9} Watts
Microwatts (μW)	10^{-6} Watts
Milliwatts (mW)	10^{-3} Watts
Watts (W)	1 Watt
Kilowatts (kW)	10^3 Watts
Megawatts (MW)	10^6 Watts

The two lowest-level subunits are used mostly in radio astronomy and other very weak signal applications, whereas the two highest-level subunits are used in powerful radio and television transmitters.

Decibel Notation of Power Units

It is common practice to express power relationships in terms of *decibel notation* (dB). This approach allows gains and losses to be added and subtracted, rather than multiplied and divided, somewhat simplifying the arithmetic.

Relative Power Levels:

$$dB = 10 \, Log\left[\frac{P1}{P2}\right]$$ [6.4]

Absolute Power Levels (50-Ω Load):

$$dBm = 10 \, LOG\left[\frac{P_W}{0.001}\right]$$ [6.5]

or,

$$dBm = 10\ LOG\ P_{mW} \qquad [6.6]$$

Where:

dBm is power level relative to one-milliwatt in a
50-Ω load
P1 and P2 are two power levels (same units)
P_W is power in watts (W)
P_{mW} is power in milliwatts (mW)

Table 6.1 shows the power levels in watts (W) and milliwatts (mW) for dBm values from −60 dBm to +60 dBm.

Types of RF Power Measurement

When you measure electrical power it helps to know just what you are measuring. In DC circuits involving only resistance (Figure 6.1A) it is straightforward: measure the voltage (V) and current (A) and find their product per Equation 6.1.

TABLE 6.1   dBm, watts, and milliwatts compared.

dBm	Power (W)	Power (mW)
− 60	1.00E − 09	1.00E − 06
− 55	3.16E − 09	3.16E − 06
− 50	1.00E − 08	1.00E − 05
− 45	3.16E − 08	3.16E − 05
− 40	1.00E − 07	1.00E − 04
− 35	3.16E − 07	3.16E − 04
− 30	1.00E − 06	1.00E − 03
− 25	3.16E − 06	3.16E − 03
− 20	1.00E − 05	1.00E − 02
− 15	3.16E − 05	3.16E − 02
− 10	1.00E − 04	1.00E − 01
− 5	3.16E − 04	3.16E − 01
0	1.00E − 03	1.00E + 00
5	3.16E − 03	3.16E + 00
10	1.00E − 02	1.00E + 01
15	3.16E − 02	3.16E + 01
20	1.00E − 01	1.00E + 02
25	3.16E − 01	3.16E + 02
30	1.00E + 00	1.00E + 03
35	3.16E + 00	3.16E + 03
40	1.00E + 01	1.00E + 04
45	3.16E + 01	3.16E + 04
50	1.00E + 02	1.00E + 05
55	3.16E + 02	3.16E + 05
60	1.00E + 03	1.00E + 06

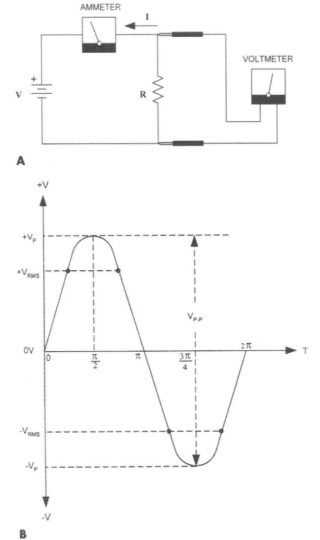

Figure 6.1 (A) RF power can be measured by measuring the RF current and load resistance and then applying I^2R; (B) Relationships found on a sine wave RF signal; (C) Phase difference between current and voltage waveforms.

AC Power

In AC circuits the problem is a little less straightforward. Figure 6.1B shows an AC sine wave. Note that over one cycle the voltage rises from zero to a positive peak value ($+V_P$) and then falls back to zero. The current follows a similar path. The current then reverses direction, rises to a negative peak ($-V_P$), and falls back to zero. What voltage is used? If we want *peak power* (rarely), then we would use the peak voltage and peak current in Equation 6.1. But the peak power is not terribly useful. We can define the *real power* in an AC circuit as the *equivalent DC power that would produce the same amount of heating in a resistive load as the applied AC waveform.* If the load is resistive, that is, there are no capacitive reactance (X_C) or inductive reactance

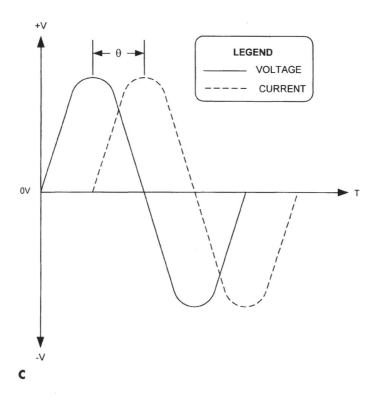

C

(X_L) elements present, then we can use the *root mean square* (RMS) values (V_{RMS} and I_{RMS}) to find this equivalent or RMS power.

$$P = V_{RMS} \times I_{RMS} \qquad [6.7]$$

Where:

$$V_{RMS} = \frac{V_P}{\sqrt{2}} \qquad [6.8]$$

$$I_{RMS} = \frac{I_P}{\sqrt{2}} \qquad [6.9]$$

The situation becomes more complex, however, when reactance is found in the circuit. In a reactive circuit there is a *phase shift* (θ) between the voltage and current. This makes the *apparent power* and real power different by the cosine of the phase angle, which is called the *power factor.*

$$P = V \times I \times Cos\ \theta \qquad\qquad [6.10]$$

When $\theta \neq 0$, we usually use the units volt-amperes rather than watts.

RF Power

Radio frequency power is essentially AC power, but certain additional problems present themselves. For a continuous wave (CW) signal (Figure 6.2A), the issue is relatively straightforward because the signal is a series of equal amplitude sine waves not unlike those of Figure 6.1B. For on-off telegraphy (Figure 6.2B), the problem gets somewhat more difficult because the waves are not of constant amplitude. In an amplitude modulated (AM) wave (Figure 6.2C), the amplitude varies with the modulating audio signal, presenting a further complication.

Pulse Parameters

The pulse signal (Figure 6.2D) may or may not vary in a regular manner. The pulse might also have very high overshoot peaks that add little to overall power, so an equation based on the peak value is largely useless.

Several pulse parameters are of interest. The *rise-time* (τ_r) of the pulse is the time required for the pulse to rise from 10% of the peak amplitude to 90% of the peak amplitude. The pulse width (τ) is the time between the 50% amplitude points. The pulse train has a period of zero amplitude between successive pulses, which is called the *off-time* (τ_{off}) The total duration of the pulse, including τ and τ_{off}, is called the *pulse repetition interval* (PRI). The pulse repetition frequency is the number of pulses per second, or

$$PRF = \frac{1}{PRI} \qquad\qquad [6.11]$$

The *duty cycle* of the pulse is the ratio of on-time (pulse width) to off-time, expressed as a percentage or:

$$Duty\ Cycle = \frac{\tau}{PRI} \times 100\% \qquad\qquad [6.12]$$

The *average power* is the energy transfer averaged over the entire pulse, including on-time and off-time. Mathematically, the power in a pulse is:

$$P_{Pulse} = \frac{1}{\tau} \int v(t)\ i(t)\quad dt \qquad\qquad [6.13]$$

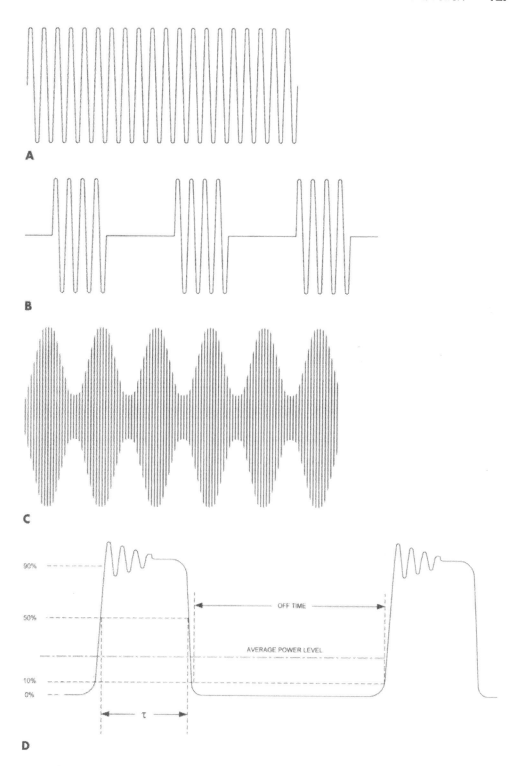

Figure 6.2 (A) CW sine wave signal; (B) Pulsed sine wave signal; (C) Amplitude-modulated signal (D) Pulse signal (with ringing) relationships.

The pulse power is also sometimes expressed in terms of:

$$P_{Pulse} = \frac{P_{Average}}{Duty\ Cycle} \qquad\qquad [6.14]$$

These equations deal with the power in one pulse. To find the power in a wave train it is necessary to multiply its amplitude by the number of pulses per second, the *pulse repetition frequency* (PRF). Table 6.2 shows the power relationships for assorted waveforms, the spectra for those waveforms, and expected readings. The figures are arbitrarily based on the peak envelope voltage (PEV) in each waveform. Also shown are the expected power readings compared with 100-watts CW power.

TABLE 6.2 Power levels associated with various waveforms. (Courtesy of Bird Electronics Corporation)

Transmission Type and Scope Pattern	Frequency Spectrum (C × Carrier)	PEV_RMS (arbitrary)	PEP = PEV²_RMS / Z₀	P_AVE (Average Heating Power)	Models 4314B, 4391A CW Mode	PEP% Mode	MOD Mode	Model 43 4304A, 4308	Model APM-16
Table A Multiple Carriers		$\frac{400}{\sqrt{2}}$ V	1600W	400W	—	1600W	—	—	400W
Table B CW		$\frac{100}{\sqrt{2}}$ V	100W	100W	100W	100W	0%	100W	100W
Table C AM 100% Mod.		$\frac{200}{\sqrt{2}}$ V	400W	150W	100W	400W	100%	100W	150W
Table D AM 75% Mod.		$\frac{173}{\sqrt{2}}$ V	300W	127W	100W	300W	73%	100W	127W
Table E SSB 1 Tone		$\frac{100}{\sqrt{2}}$ V	100W	100W	100W	100W	0%	100W	100W
Table F SSB 2 Tones		$\frac{100}{\sqrt{2}}$ V	100W	50W	25W	100W	100%	40.5W	50W
Table G SSB Voice		$\frac{100}{\sqrt{2}}$ V	100W	—	—	100W	—	—	—
Table H TV Black Level		$\frac{100}{\sqrt{2}}$ V	100W	60.1W	Models 4314B, 4391A only —	100W	—	59.6W	60.1W
Table 1 Pulse		$\frac{100}{\sqrt{2}}$ V	100W	10W	—	100W	100%	—	10W

METHODS FOR MEASURING RF POWER

RF power meters use a number of different approaches to making the measurement. Some will measure the current or the voltage at a resistive load. Others are based on the fact that power dissipated in a resistive load is given off as heat, so the temperature change can be used as the indicator of RF power. This approach has the advantage of finding a DC equivalent power.

Figure 6.3 shows the basic scheme. A load resistor, R_O, with a resistance equal to the system impedance, is enclosed in an isolated environment with some sort of temperature sensor. Theoretically, one could place a dummy load resistor in a room, and then use a glass mercury thermometer and stopwatch to measure the rise in temperature and elapsed time to find the power. That's hardly practical, however. The basic idea is to find a transducer that will convert the heat generated in the load resistor to either a DC or low-frequency AC signal that is easily measured with ordinary electronic instruments.

Thermistor RF Power Meters

A *thermistor* is a resistor that changes its electrical resistance with changes in temperature. Although all conductors exhibit some "thermistor behavior," actual thermistors are usually made of a metallic oxide compound. Figure 6.4A shows the resistance-versus-temperature curve for a typical thermistor device. A *negative temperature coefficient* (NTC) device will decrease resistance with increases in temperature. A positive *temperature coefficient device* (PTC) device does the opposite: resistance rises with increases in temperature.

Bolometers

Figure 6.4B shows a family of resistance versus self-heating power curves for a single thermistor operated at different temperatures. The resistance is not only nonlinear, which makes measurements difficult enough in its own right, the shape and placement of the curve vary with temperature. As a result, straight thermistor instruments can be misleading. *Bolometry* is a method that takes advantage of this problem to create a more accurate RF power measurement system.

Self-heating power is caused by a DC bias flowing in the thermistor. Figure 6.4C shows how self-heating can be used in bolometry. The thermistor (RT1) is adjusted to a specified self-heating point when no RF power is applied to the dummy

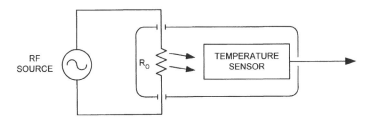

Figure 6.3 Simple bolometry/calorimetry approach to measuring RF power.

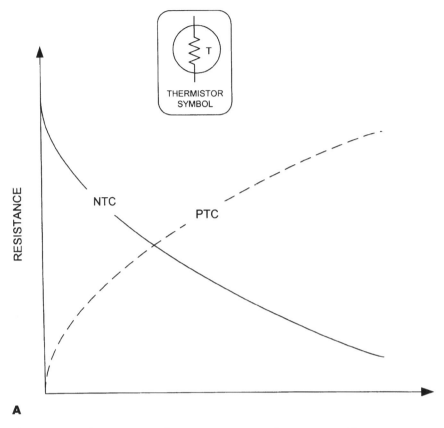

A

Figure 6.4 A) Thermistor R-vs.-T characteristics; (B) R-vs.-P characteristics at different ambient temperatures; (C) Thermistor sensor mount for measuring RF power.

load R_O. The resistance of thermistor RT1 can be read from the voltmeter because the current from the constant current source remains the same once it is adjusted to a set point.

When RF power is applied to the dummy load, heat radiated from the load will cause the resistance of RT1 to decrease. The bolometer current source is then adjusted to decrease the bias until the resistance rises back to the value it had before power was applied. This point is indicated by returning the meter reading to the same point as before. The change of bias power required to restore the thermistor to the same resistance is therefore equal to the power dissipated in the dummy load.

Self-Balancing Bridge Instruments

The *Wheatstone bridge* circuit (Figure 6.5) is used in a number of instrumentation circuits. In the null condition ($V_O = 0$) the ratios of the resistors are equal: R1/R2 = R3/R4. It is not strictly necessary that R1 = R2 = R3 = R4, only that R1/R2 = R3/R4, but the condition R1 = R2 = R3 = R4 = R is often the case. If one of the resistors is placed with a thermistor, then the temperature can be measured

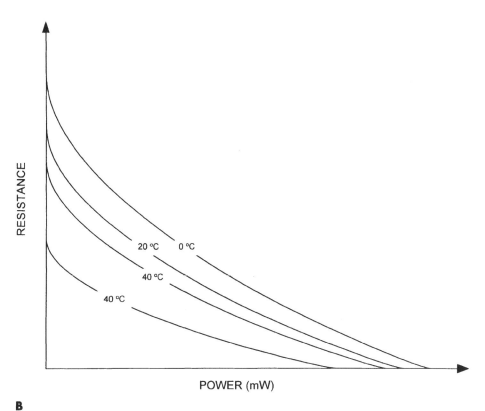

B

C

by the imbalance of the bridge. Similarly, if the thermistor resistance is such that the equality R1/R2 = R3/R4 is satisfied, then we can infer the resistance of the thermistor by the null condition.

The self-balancing (i.e., autobalancing or autonull) bridge shown in Figure 6.6 uses a Wheatstone bridge thermistor to perform bolometry measurement of RF

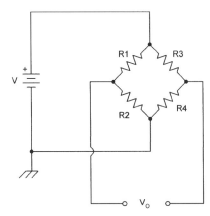

Figure 6.5
Wheatstone bridge circuit with a thermistor in one arm.

power. The thermistor mount-sensor assembly contains a dummy load and a thermistor (R_T). The null condition is created when $R1/R_T = R2/R3$.

The self-balancing bridge uses a differential amplifier (A1) to perform the balancing. A differential amplifier produces an output voltage proportional to the difference in two input voltages. When the Wheatstone bridge is in balance, then the output of the differential amplifier is zero. The bias for the Wheatstone bridge, hence the thermistor in the bolometer sensor, is derived from the output of the amplifier. A change in the resistance of the thermistor unbalances the bridge, and this moves the amplifier's differential input voltage away from zero. The amplifier output voltage goes up, thereby changing the bias current in an amount and direction necessary to restore balance. Thus, by reading the bias current we can infer the power level that changed the thermistor resistance.

Because the thermistor will have a different characteristic curve at different ambient temperatures, it is necessary either to control the ambient temperature or to correct for it. It is very difficult to control the ambient temperature. Although it

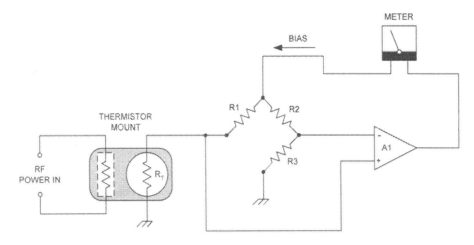

Figure 6.6 Self-nulling or self-balancing Wheatstone bridge circuit.

is done, it is also not terribly practical in most cases. As a result, it is common to find RF power meters using two thermistors in the measurement process (Figure 6.7). One thermistor is mounted in the thermistor sensor mount, while the other is used to measure the ambient temperature. The readings of the ambient thermistor are used to correct the readings of the sensor thermistor.

Thermocouple RF Power Meters

The thermocouple is one of the oldest forms of temperature sensor. The basis for the thermocouple is what happens when a metal wire is heated (Figure 6.8). The heated end produces a larger number of free electrons due to thermal agitation. There is a gradient of free electrons from the hot end to the cold end because of the temperature difference. As each electron migrates away from the hot end, a positive ion ("hole") is created that tends to drag the electrons back towards the left. The leftward drift produces an electric field that serves as a voltage source.

This effect differs from one metal to another. When two *dissimilar metals* are connected together to form a junction (Figure 6.9), and the junction is heated, then the potential across the free ends (V_T) is proportional to the temperature of the hot junction. This phenomenon is called the *Seebeck effect*. An approximation of the output voltage is given by:

$$V_T = a + bT + cT^2 + dT^3 + eT^4 + fT^5 \qquad [6.15]$$

Where:

V_T is the output voltage

T is the applied temperature

a, b, c, d, e, f . . . are coefficients that depend on the nature of the particular junction.

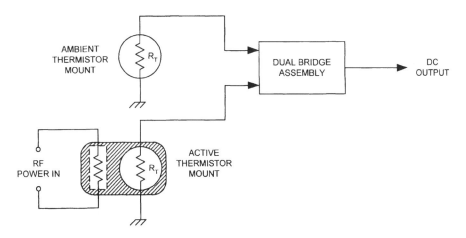

Figure 6.7 Method for accounting for ambient temperature variation in bridge assembly.

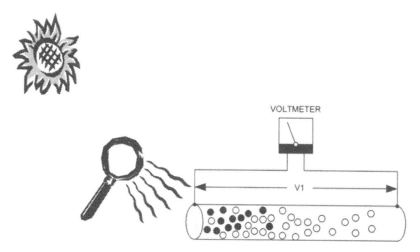

Figure 6.8 Thermocouple action depends on temperature effects in conductors affecting charge distribution.

RF Ammeters

RF ammeters are constructed using thermocouples and a small value resistance heating element (Figure 6.10). The meter will have a small wire resistance element in close proximity to a thermocouple element. The thermocouple is, in turn, connected to a DC meter. When current flows through the resistance-heating element, the potential across the ends of the thermocouple changes proportional to the RMS value of the current. Thus, the RF ammeter measures the RMS value of the RF current.

If the RF ammeter is used to measure the current flowing from an RF source to a resistive load, then the product I^2R indicates the true RF power. These meters can measure current, and infer power from I^2R up to 50 or 60 MHz, depending on the instrument.

Thermocouples and thermistors share the ability to measure true RF power. Although thermocouple RF ammeters have been used since the 1930s, or earlier,

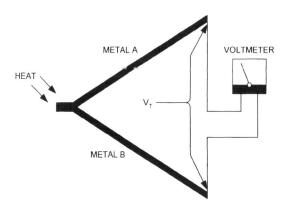

Figure 6.9
Bimetallic thermocouple produces an output voltage proportional to the applied temperature and the work function difference between the two metals.

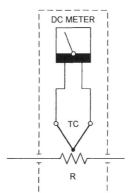

Figure 6.10
Thermocouple RF ammeter.

the use of thermocouples in higher-frequency and microwave power meters started in the 1970s. Thermocouples are more sensitive than thermistor sensors and are inherently square law devices.

Figure 6.11 shows a solid-state thermocouple sensor that is used well into the microwave region. Two semiconductor thermocouples are connected such that they are in series for DC, and parallel for RF frequencies. Thus, their combined output voltages are read on the DC voltmeter. Because of the capacitors, however, they are in parallel for RF frequencies, and if designed correctly will make a 50-Ω termination for a transmission line.

Thermocouples suffer the same reliance on knowing the ambient temperature as thermistors. Figure 6.12 shows a method for overcoming this problem. A pair of thermistors is used. One is used either in a bolometry circuit or as a terminating sensor to measure the unknown RF power. The other sensor is used to measure a highly controlled reference power source. Depending on the implementation, the reference power might be DC or another RF oscillator with a highly controlled, accurately calibrated output power level.

Diode Detector RF Power Meters

Rectifying diodes convert bi-directional alternating current to unidirectional pulsating DC. When filtered, the output side of a diode is a DC level that is proportional to the amplitude of the applied AC signal. Figure 6.13 shows the unidirectional action in the form of the *I-versus-V* curve. When the applied bias is positive ("forward bias") the current will begin to flow, but not proportionally. At some point (600 to 700 mV in silicon diodes), marked as Vγ in Figure 6.13, the response enters a linear region. This response is termed *ohmic* because it follows Ohm's law.

When the applied voltage reverse biases the diode, the current flow ceases, except for a very small *leakage current* (I_L). One indicator of the quality of diodes is that I_L is minimized on higher-quality units.

The nonlinear region of the I-versus-V curve is called the *square law region*. In this region the rectified output voltage from the diode is proportional to the input power (Figure 6.14). This behavior is seen from power levels of −70 to −20 dBm.

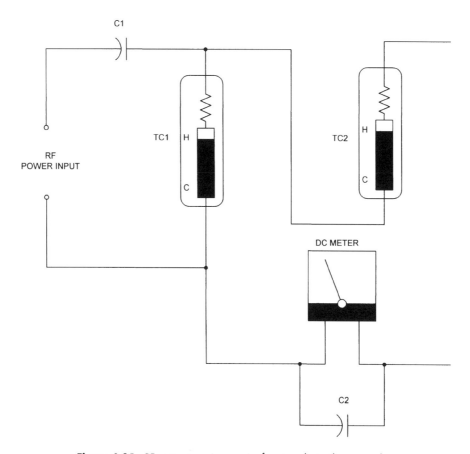

Figure 6.11 RF wattmeter using a pair of semiconductor thermocouples.

In low-cost RF power measuring instruments silicon and even germanium diodes are often used, but these are not highly regarded for professional measurements. Low-barrier Schottky diodes are widely used well into the microwave region. For higher frequencies in the microwave region, planar doped barrier (PDB) diodes are preferred. They work up to 18 GHz or better, and with power levels of −70 dBm. It is claimed that PDB diodes are more than 3,000 times more efficient than thermocouple detectors.

Circuits

Figures 6.15 and 6.16 show two similar circuits using a diode detector. Resistor R_L in Figure 6.15 is a *dummy load* that has a resistance value equal to the characteristic impedance of the transmission line connecting the system (e.g., 50 ohms). Diode D1 is the rectifier diode, while capacitor C1 is used to filter the pulsations at the rectifier output into pure DC. A problem with that circuit is that it is limited to power levels consistent with the native characteristics of the diode.

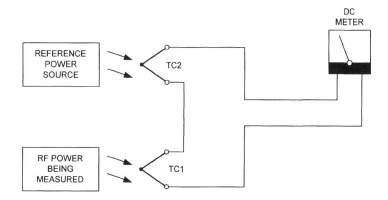

Figure 6.12 Using two thermocouples to compare unknown RF power with a known DC or low frequency AC power source.

Higher Power Circuit

Figure 6.16 shows the same circuit with a resistor voltage divider (R1/R2) to reduce the voltage associated with higher power levels to the characteristic of the diode. The actual voltage applied to the diode will be $V_{RL} \times [R2/(R1 + R2)]$.

Low Power Levels

At very low power levels the diode output voltage drops very low. For example, at a power level of −70 dBm, the diode produces about 50 nanovolts (nV) output

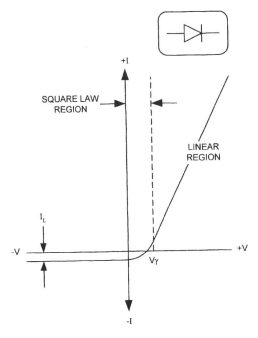

Figure 6.13
I-vs.-V curve of diode rectifier.

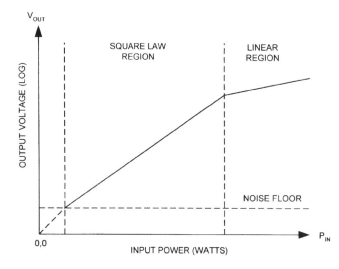

Figure 6.14 Output voltage vs. RF input power curve.

potential. This level is too close to the noise and the drift values of typical DC amplifiers to be of use. A solution is to use a *chopper circuit* as shown in Figure 6.17.

A chopper is an electronic switch that turns the DC signal from the diode output on and off at a high rate (typically 100 to 10,000 times per second). The switching action is created by either a square wave generator or sine wave "carrier" oscillator that is applied to the toggle input of the electronic switch.

The chopped signal is essentially an AC signal, so it can be amplified in an AC amplifier, which has a much smaller feedback controlled drift figure than do DC amplifiers. Also, the AC signal can be bandpass-filtered to remove noise. The bandpass filter is centered on the frequency of the carrier oscillator.

The chopped, amplified, and filtered signal is applied to a synchronous detector that is controlled by the same carrier oscillator that performed the chopping action. A low-pass filter following the synchronous detector removes residual components of the switching action at the carrier frequency. Finally, a DC amplifier provides scaling to the correct DC level, or as level translation for an analog-to-digital converter (ADC).

The Thruline® Sensor

The Bird Electronics (30303 Aurora Road, Cleveland, OH, 44139; 1-440-248-1200) Thruline® sensor is shown in Figure 6.18A, and an equivalent circuit is shown in

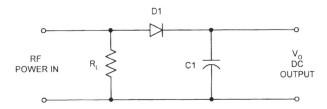

Figure 6.15 Simple diode detector for measuring RF power.

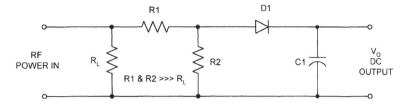

Figure 6.16 Simple diode detector circuit for higher power levels.

Figure 6.18B. The sensor consists of a coaxial transmission line section and a wire loop directional coupler that connects to a diode detector (D1).

Consider the equivalent circuit in Figure 6.18B. The factor M is the mutual coupling between the loop and the center conductor of the coaxial line section, as well as the voltage divider consisting of R1 and C1. Potential E is the voltage between the inner and outer conductors of the coaxial line, while E_R is the voltage drop across the resistor; e_M is the voltage across the inductor, while e is the output potential. We know that the R1-C1 voltage divider produces a potential given by Equation 6.16, provided that $R \ll X_C$ and $e_m = Ij\omega \pm M$.

$$e_R = \frac{RE}{X_C} = R\,E\,j\,\omega\,C \qquad\qquad [6.16]$$

The output voltage is:

$$e = e_R + e_M = j\omega(CRE \pm MI) \qquad\qquad [6.17]$$

The values of the components are selected such that $R \ll X_C$ and $CR = M/Z_0$. We can now state that the DC output voltage is:

$$e = j\omega\left[\frac{EM}{Z_o} \pm MI\right] = j\omega M(E/Z_o \pm I) \qquad\qquad [6.18]$$

At any point along a transmission line the voltage appearing between the center conductor and outer conductor (E) is a function of the forward voltage (E_F) and the reflected voltage (E_R). When the directional coupler is pointed at the load, the output voltage of the sensor reads the forward voltage and produces an output voltage of

$$e = \frac{j\omega M\,E_F}{Z_o} \qquad\qquad [6.19]$$

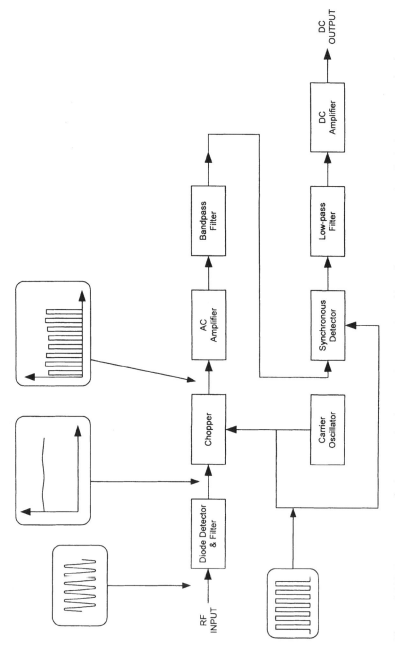

Figure 6.17 RF power meter for low power levels uses a chopper and amplifier scheme to measure low diode detector output levels.

And when pointed at the source:

$$e = \frac{j\omega M \, E_R}{Z_o} \qquad\qquad [6.20]$$

Thus, this sensor produces a voltage that is a function of the direction of the RF signal flowing in the transmission line.

Figure 6.19 shows two examples of the Thruline® instruments by Bird Electronics. The instrument in Figure 6.19A is the Model 4410A. It is based on the classic Model 43 design. It offers an insertion VSWR of 1.05:1 up to 1,000 MHz. The sensor elements are plug-in and have an arrow to indicate the direction of the

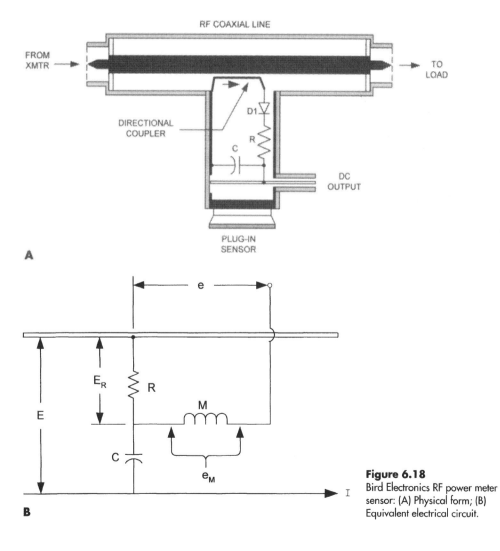

Figure 6.18
Bird Electronics RF power meter sensor: (A) Physical form; (B) Equivalent electrical circuit.

measurement (pointed towards the load or the source, depending on whether you measure P_F or P_R). One difference between this instrument and earlier instruments is that there is a factor control on the meter to optimize performance for the specific element inserted.

Another variation on the theme is shown in Figure 6.19B. This instrument has a digital readout and is microprocessor controlled. Two elements are used to measure P_F and P_R simultaneously. Also shown in Figure 6.19B is the IEEE-488 GPIB (General Purpose Interface Bus) interface box. This accessory allows remote operation through a computer network, and perhaps more importantly, it allows the meter to be incorporated in automatic test equipment systems.

Figure 6.19C shows the characteristic of several Series-C plug-in elements. These elements are selected according to the power level and frequency range anticipated.

Calorimeters

Calorimeters are capable of making very accurate measurements of RF power, especially at high power levels where other methods tend to fall down. These in-

A

Figure 6.19
(A) Bird Model 4410A RF wattmeter; (B) Bird RF Power Analyst; (C) Bird RF power meter element frequency response curves. (Photos and chart courtesy of Bird Electronics Corporation)

B

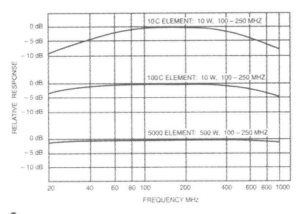

C

struments measure the heating capability of the RF waveform, so they produce an output proportional to the RMS power level that is independent of the applied waveform. The First Law of Thermodynamics is the basis for the operation of calorimeters: *energy can neither be created nor destroyed, only changed in form.* There are two basic forms of calorimeter: dry and flow (or wet). The dry calorimeters are used at lower power levels and are represented by the thermistor and thermocouple methods discussed earlier in this chapter. Flow calorimeters are used at higher power levels.

Flow calorimeters also come in two varieties: *substitution flow* and *absolute flow*. Power can be measured using the following relationship:

$$P = F_{mass} \times (T_{OUT} - T_{IN}) \times C_P(T) \qquad [6.21]$$

Where:

P is the power level

F_{Mass} is the mass flow rate of the fluid used in the calorimeter

T_{OUT} is the fluid temperature after being heated by the RF load resistor

T_{IN} is the fluid temperature before being heated by the RF load resistor

$C_p(T)$ is the fluid specific heat as a function of temperature (T)

Substitution Flow Calorimeters

This form of RF power meter (Figure 6.20) uses two fluid loops. Each fluid loop is heated by a separate termination resistor. Termination "A" is heated by a low frequency AC power source, and the power applied to this termination is measured by an AC power meter. The unknown RF power is applied to Termination "B." The differential temperature (T_{OUT}-T_{IN}) is measured by a differential thermocouple.

When the temperatures of the two fluids are equal to each other, then the output of the thermocouple is zero. When the AC power is adjusted to balance the temperatures while RF is applied, producing a zero output voltage from the thermocouple, the RF power is equal to the more easily measured AC power. A temperature stabilizer and heat exchanger returns the temperature of the fluid to base level after it is used to measure power.

This method will produce error of 0.28% or better, up to RF power levels of one kilowatt. Both water and oil are used as fluid coolants in various instruments.

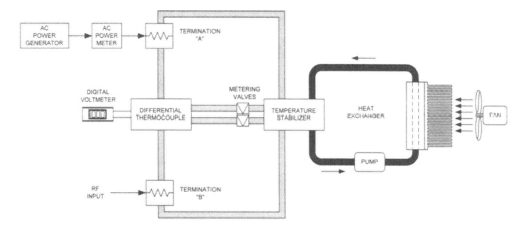

Figure 6.20   Substitution flow calorimeter.

Absolute Flow Calorimeters

Figure 6.21 shows the absolute flow calorimeter. This type of RF power meter measures the mass flow rate of the coolant, as well as the temperatures before (T_{IN}) and after (T_{OUT}) the RF load resistor. The mass flow rate is:

$$F_{Mass} = f W_S (T_{IN}) \qquad\qquad [6.22]$$

Where:

W$_S$ is the specific weight of the fluid at the input temperature
f is the volume flow rate (liters/min)
All other terms as previously defined

By combining Equations 6.21 and 6.22, we get the equation for measuring RF power by this means:

$$P = k \times f \times W_S(T_{IN}) \times C_P(T_{AVE}) \times (T_{OUT} - T_{IN}) \qquad [6.23]$$

Where: $T_{AVE} = (T_{OUT} - T_{IN})/2$ and all other terms are as previously defined.

One of the advantages of the absolute flow approach is that it does not depend on nulling or calibration of a low-frequency power source, yet it produces good accuracy at high power levels (up to 80 kW). Figure 6.22 shows a commercially available calorimeter RF power meter.

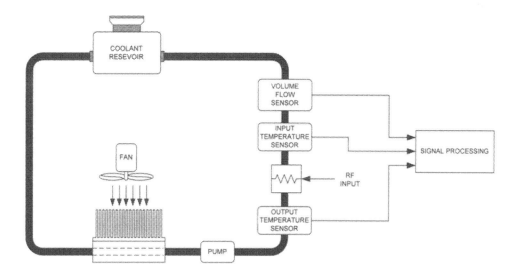

Figure 6.21 Absolute flow calorimeter.

Figure 6.22 Bird calorimeter RF wattmeter. (Photo courtesy of Bird Electronics Corporation)

MICROPOWER MEASUREMENTS

Micropower measurements pose special problems because they are made at levels below the range of most practical RF power sensors. In some cases, the chopper approach can be used with a diode detector. At lower levels, however, some other method is needed. Figure 6.23 shows a comparison method using a calibrated RF signal generator. The instrument selected must have a calibrated output attenuator that provides accurate outputs in either dBm or microvolts (µV). The signal generator and the unknown micropower source are connected to a receiver equipped with an S-meter through a hybrid coupler. The coupler must have either equal port-to-port losses for the two inputs, or at least accurately known different losses. Optional calibrated step attenuators are also sometimes used to balance the power levels. The receiver acts as a micropower wattmeter or voltmeter because it will produce an S-meter reading of even very weak signals.

Two methods can be used: either *equal deflection* or *double deflection.* In the equal deflection method, the unknown source is turned on and the S-meter reading noted. The unknown source is then turned off, and the signal generator is turned on. The output of the signal generator is adjusted to produce the same

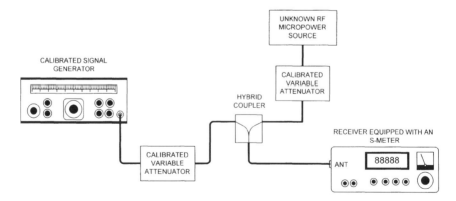

Figure 6.23 Method for measuring micropower RF levels with some degree of accuracy.

S-meter deflection. The power level of the unknown source is therefore equal to the calibrated signal generator output level.

The double deflection method sets the signal generator output to zero and then applies the unknown RF power to the receiver. The S-meter reading is noted (for practical reasons, adjust the attenuator to let the meter fall on a specific indicator marking). The signal generator output is then increased until the S-meter reading goes up one S-unit (which will be either 3 or 6 dB, depending on the design of the receiver). The output level of the signal generator is therefore equal to the unknown power source.

Error and Uncertainty Sources

All measurements have some basic error, that is, a difference between the actual value of a variable and the value read from a meter. The three dominant classes of error in RF power measurements are *mismatch uncertainty, sensor uncertainty,* and *meter uncertainty.*

Meter uncertainty is error due to problems in the meter-indicating device itself. It might be a measurement error, such as a difference between the actual output voltage and the displayed output voltage (which represents power). You might see zero set error, zero carryover, drift, noise, and other sources of instrument error.

On analog meters there are also additional error sources. For example, the width of the pointer covers a certain distance on the scale, so it creates a bit of ambiguity. Also, there may be a parallax error if the meter is not read straight.

Digital meters exhibit quantization error and last digit bobble error. The quantization error comes from the fact that the digital representation of a value can only assume certain discrete values, and an actual value might be halfway between the two authorized levels. Last digit bobble (± 1 count) error results from the fact that the least significant digit tends to bounce back and forth between two adjacent values.

Sensor error may come in a variety of guises, depending on the nature of the sensor. Thermistors and thermocouples, for example, have different forms of error. Most sensors, though, exhibit an effective efficiency error due to losses in the sensor. It occurs when some of the applied RF energy is radiated as heat rather than being used to affect the output reading. This problem may be expressed as a calibration uncertainty or calibration factor by the manufacturer of the sensor.

Mismatch Loss and Mismatch Uncertainty

The mismatch loss occurs when a standing wave ratio (SWR or VSWR) exists in the system. Maximum power transfer occurs when a source impedance and a load impedance are matched (Figure 6.24). If these impedances are not matched, then a portion of the power sent from the source to the load is reflected. The *reflection coefficient* (ρ) is:

$$\rho = \frac{VSWR - 1}{VSWR + 1}$$

[6.24]

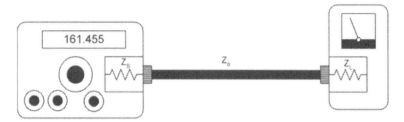

Figure 6.24   Direct measurement of RF power requires impedance matching throughout system, otherwise errors will result.

Table 6.3 shows the reflection coefficient for VSWR values from 1:1 to 3:1. The
single-ended *mismatch loss* in decibels is

$$M.L. = 10 \log \left[1 \pm \rho^2 \right] \text{dB} \qquad [6.25]$$

If the system is mismatched on both ends (Figure 6.24), then mismatch
loss is:

$$M.L. = 20 \log \left[1 \pm (\rho_1 \times \rho_2) \right] \qquad [6.26]$$

The mismatch uncertainty, expressed as a percent, is:

$$M.U. = \pm 2 \times \rho_1 \times \rho_2 \times 100\% \qquad [6.27]$$

Let's assume that there is a 1.75:1 VSWR at the source end ($\rho 1 = 0.27$) and a
VSWR of 1.15:1 ($\rho 2 = 0.07$) at the sensor/load end. The mismatch uncertainty is:

$$M.U. = \pm 2 \times 0.27 \times 0.07 \times 100\% = 3.78\% \qquad [6.28]$$

Total Uncertainty

The total uncertainty in the measurement involves the mismatch uncertainty, calibration factor uncertainty, instrumentation uncertainty, and power source uncertainty (if a reference power source is used in a comparison measurement). There
are several ways to state the total uncertainty, two of which are the *worst case uncertainty* and *root sum square uncertainty*. The worst case uncertainty is the sum of

TABLE 6.3   VSWR and associated reflection coefficients and single ended loss (in decibels).

VSWR	Refl Coeff.	Single Loss (dB)
1.00	0.00	0.000
1.05	0.02	0.003
1.10	0.05	0.010
1.15	0.07	0.021
1.20	0.09	0.036
1.25	0.11	0.054
1.30	0.13	0.075
1.35	0.15	0.097
1.40	0.17	0.122
1.45	0.18	0.149
1.50	0.20	0.177
1.55	0.22	0.207
1.60	0.23	0.238
1.65	0.25	0.269
1.70	0.26	0.302
1.75	0.27	0.336
1.80	0.29	0.370
1.85	0.30	0.405
1.90	0.31	0.440
1.95	0.32	0.475
2.00	0.33	0.512
2.05	0.34	0.548
2.10	0.35	0.584
2.15	0.37	0.621
2.20	0.38	0.658
2.25	0.38	0.695
2.30	0.39	0.732
2.35	0.40	0.770
2.40	0.41	0.807
2.45	0.42	0.844
2.50	0.43	0.881
2.55	0.44	0.919
2.60	0.44	0.956
2.65	0.45	0.993
2.70	0.46	1.030
2.75	0.47	1.067
2.80	0.47	1.103
2.85	0.48	1.140
2.90	0.49	1.177
2.95	0.49	1.213
3.00	0.50	1.249

all individual uncertainties in the direction that maximizes the overall uncertainty. For example, suppose we have a system where:

Mismatch uncertainty	3.78%
Calibration factor uncertainty	1.76%
Instrumentation uncertainty	0.95%
Power reference uncertainty	1.35%

The worst case uncertainty is their sum:

$$\text{Uncertainty} = \pm(3.78\% + 1.76\% + 0.95\% + 1.35\%) = \pm 7.84\%$$

Real errors are rarely worst case, but rather are uncorrelated to each other. The root sum squares (RSS) method allows a single error term to represent the average errors of the system. For a system with four sources of error (as above)—E1, E2, E3, E4, and E5—the RSS error is:

$$RSS = \sqrt{E1^2 + E2^2 + E3^2 + E4^2} \qquad [6.29]$$

In terms of the values above:

$$RSS = \sqrt{(3.78\%)^2 + (1.76\%)^2 + (0.95\%)^2 + (1.35\%)^2}$$
$$RSS = \sqrt{14.29 + 3.1 + 0.90 + 1.82}\ \% = \sqrt{20.11}\ \% = 4.48\%$$

Compare the worst case error of 7.84% with the RSS error of 4.48%.

Expressed in terms of decibels, the RSS percent loss is:

$$RSS(dB) = 10\log\left[1 \pm (RSS\%/100)\right] \qquad [6.30]$$

CHAPTER SEVEN

Measuring Frequency and Period

The measurement of the frequency and period of signal waveforms is often required in RF systems. Frequency is especially important because the tuning accuracy of receivers, and the output frequency of transmitters, must be properly set or the system will not work properly.

THE BASICS

Although a number of waveforms are sometimes encountered, let's start our discussion with the basic sine wave (Figure 7.1). The *cycle* is one complete positive and negative excursion of the sinusoidal AC waveform. It can be described by comparing identical features on successive waveforms, as shown in Figure 7.1. The *frequency* (F) of the waveform is the number of cycles that occur in one second (1 s). Frequency is measured in Hertz (Hz), in which one hertz equals one cycle per second, or the superunits kilohertz (thousands of Hz) and megahertz (millions of Hz). The *period* of the waveform is the length of time (T) required for one entire cycle to occur. The period and frequency are reciprocals of each other (Equation 7.1), so it is not surprising that the measurement schemes are similar.

$$F = \frac{1}{T} \tag{7.1}$$

Where:

    F is the frequency in Hertz (Hz)
    T is period in seconds (s)

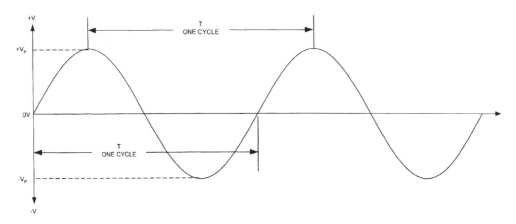

Figure 7.1 Frequency and period compared on a sine wave.

Rough Measurement Methods

It is sometimes necessary to make very rough, nonprecise measurements of frequency. Such a case might be when testing an RF power amplifier in a transmitter for parasitic oscillations. Because some transmitters have very high output power levels, parasitic oscillations can cause serious disruption on other channels.

Wavemeters

One way to make rough measurements of frequency is the *wavemeter* (Figure 7.2). This circuit is a descendent of wavemeters used in the 1920s, in which a tuning capacitor and inductor were connected in series with a low-voltage incandescent panel lamp. When energy from a transmitter was proximity-coupled into the coil, the lamp could be made to light up when resonance was reached. The circuit of Figure 7.2 works by rectifying the voltage across an LC-tuned circuit (L1/C1), filtering it, and then applying the resultant DC to an analog meter (M1).

Receiver

A radio receiver can also be used to measure the frequency of signals. The receiver is able to measure the frequency of relatively low-level signals.

Method 1

Adjust the receiver to the signal of interest, making sure that the signal is centered in the receiver passband. This point could be indicated by either the highest output level, or the highest S-meter reading. The frequency of the signal can then be read from the receiver's frequency dial.

This advantage is at least as good as the wavemeter method. Because receivers today tend to be rather well calibrated, the accuracy is a lot better than the wavemeter approach. The accuracy is limited primarily by the frequency accuracy of the receiver's dial and the operator's ability to center the receiver over the signal.

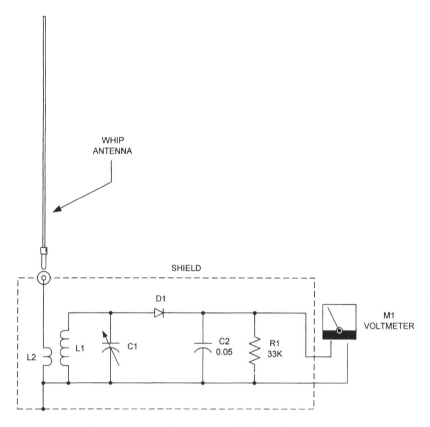

Figure 7.2 Simple wavemeter/field-strength meter circuit.

Method 2

A variation on the theme is to use an external crystal-marker generator to calibrate the receiver (Figure 7.3A). The output of the marker generator is loosely coupled to the receiver's antenna input circuit. This method was used mostly with poorly calibrated dials on analog receivers. It can be used to interpolate frequency on either analog or digital readout receivers.

Method 3

This approach is similar to the marker method but uses a variable frequency transfer oscillator or signal source (see Chapter 3, "Smith Charting the Radio Frequency Circuit") rather than a crystal-marker oscillator as an external standard. The frequency of the external standard is adjusted to the zero-beat point. This point is reached when the unknown frequency being measured is equal to the external standard frequency. The two frequencies mix together, creating a difference heterodyne. When the heterodyne frequency goes to zero, then the two frequencies are equal.

Unfortunately, the human ear can only hear down to 30 Hz or so, and that point establishes the basic accuracy of the method when done aurally. However, if the receiver is equipped with an S-meter to indicate signal strength, then as the zero

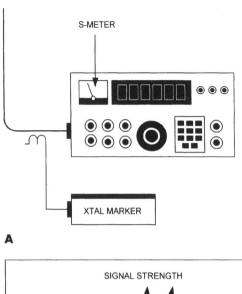

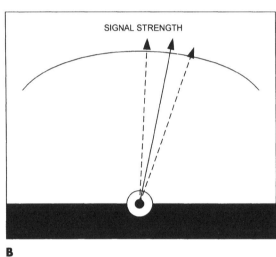

Figure 7.3
Use of a crystal marker generator
to measure frequency by
calibrating receiver dial: (A)
equipment connection; (B) Marking
zero-beat by watching S-meter
movement.

beat is approached, the meter will bob back and forth at a rate equal to the frequency difference (Figure 7.3B).

Comparison Methods

The third method of using a receiver for measuring frequency is similar to other
comparison methods.

Heterodyne Methods

Several different approaches to frequency measurement are based on the heterodyne phenomenon. If two frequencies (F1 and F2) are mixed together in a nonlinear circuit, a number of products are produced according to the rule mF1 ± nF2.

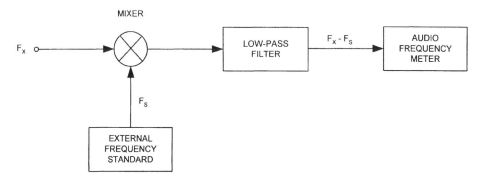

Figure 7.4 Heterodyne frequency meter block diagram.

Some of the difference products (mF1 − nF2) will fall in the audio range. If the audio frequency difference product can be brought to zero beat, and one of the frequencies is well known, then it is possible to measure the frequency of the unknown

Figure 7.4 shows a simplified block diagram for a *heterodyne frequency meter.* In this instrument, the unknown frequency F_X is mixed nonlinearly with a known signal (F_S) from a calibrated source. If F_S is precisely known, and there exists the ability either to measure the audio difference frequency accurately or detect the zero-beat point, then the unknown can be measured.

The heterodyne method, although once the mainstay of RF frequency measurement, has been largely eclipsed by digital frequency counters.

DIGITAL FREQUENCY AND PERIOD COUNTERS

Digital counters were once extremely costly, and therefore found only in the most expensive of instruments. Today, however, high-quality digital counters are available for frequency, period, and other measurements at quite reasonable costs.

Figure 7.5 shows the basic digital frequency counter, while Figure 7.6 shows a representative timing diagram. These devices are special examples of a class of instruments called *events per unit of time* (EPUT) counters. A chain of digital counters with *binary coded decimal* (BCD) outputs drives any of several kinds of digital display devices. The "events" being counted are the cycles of the input waveform. The unit of time is established by controlling a main gate, which admits signals to the counter, with a timebase.

The timing diagram of Figure 7.6 shows the action. The timebase pulses control a main gate flip-flop that changes output state on the downslope of every timebase clock pulse. The output of the main gate flip-flop is used to hold the main gate open, so as to pass pulses to the counter. If the gate time is controlled precisely, then the accumulated count will indicate the frequency of the input waveform.

Timebase Considerations

Clearly, the timebase section of the counter is critical to determining the quality of the instrument. This section consists of a crystal controlled clock oscillator and a

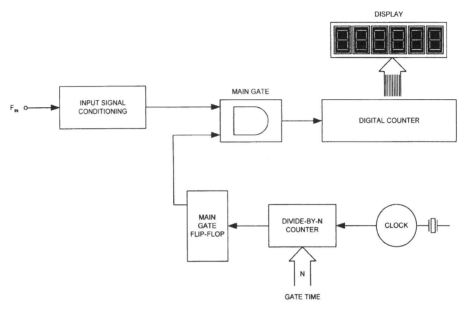

Figure 7.5 Digital frequency counter block diagram.

binary divide-by-N section. Because frequency is measured in *seconds* (s), the normal timebase circuit will open the gate for periods such as 0.01 s, 0.1 s, or 1 s. Some averaging frequency counters may use a 10-s timebase. The timebase clock oscillator is usually operated at a frequency of 1 MHz, 5 MHz, 10 MHz, or some other frequency that can be divided down to the frequency required to obtain the desired timebase gate time.

There are three basic categories of timebase crystal oscillator: *room temperature crystal oscillators* (RTXO), *temperature compensated crystal oscillators* (TCXO), and *oven controlled crystal oscillators* (OCXO).

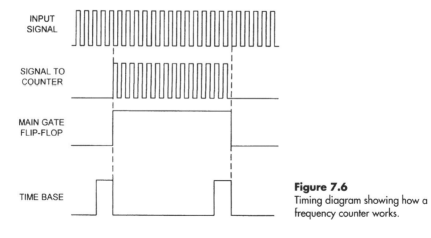

Figure 7.6
Timing diagram showing how a frequency counter works.

Room Temperature Crystal Oscillators

The RTXO takes no special precautions about frequency drift. But with proper selection of crystal cut, and reasonable attention to construction, stability on the order of 2.5 parts per million (ppm), 2.5×10^{-6}, over the temperature range 0 °C to 50 °C is possible. The RTXO is only used on economy-model counters used for non-critical applications.

Temperature Compensated Crystal Oscillators

The TCXO circuit also works over the 0 °C to 50 °C temperature range, but it is designed for much better stability. The temperature coefficients of certain components of the TCXO are designed to counter the drift of the crystal, so the overall stability is improved to 0.5 ppm (5×10^{-7}). The cost of TCXOs has decreased markedly over the years, to the point where relatively low-cost upgrades to economy counters gives them a rather respectable stability specification.

Oven Controlled Crystal Oscillators

The best stability is achieved from the OCXO timebase. These oscillators place the resonating crystal inside of a heated oven that keeps its operating temperature constant, usually near 70 °C or 80 °C.

There are two forms of crystal oven used in OCXO designs: *on/off* and *proportional control*. The on/off type is similar to the simple furnace control in houses. It has a snap action that turns the oven heater on when the internal chamber temperature drops below a certain point, and off when it rises to a certain maximum point. The proportional control type operates the heating circuit continuously, and supplies an amount of heating that is proportional to the actual temperature difference between the chamber and the set point. The on/off form of oven is capable of 0.1 ppm (10^{-7}).

OCXOs that use a proportional control oven can reach a stability of 0.0002 ppm (2×10^{-10}) with a 20-minute warm-up and 0.0001 ppm (1.4×10^{-10}) after 24 hours. It is common practice to design the counter to leave the OCTX turned on even when the counter is off. Some portable frequency counters, such as those used in two-way radio servicing, have a battery back-up to keep the OCXO turned on while the counter is in transit.

The variation described above is referred to as the *temperature stability* of the counter timebase. We must also consider *short-term stability* and *long-term stability* (aging).

Short-Term Stability

The short-term stability is the random frequency and phase variation due to noise that occurs in any oscillator circuit. It is sometimes also called either *time domain stability* or *fractional frequency deviation*. In practice, the short-term stability has to be a type of RMS (root mean square) value averaged over one second. The short-term stability measure is given as $\sigma(\Delta f/f)(t)$. Typical values of short-term stability are given below for the different forms of clock oscillator.

RTXO2 $\times$ 10^{-9} rms0.002 ppm
TCXO1 $\times$ 10^{-9} rms0.001 ppm
OCXO (on/off)5 $\times$ 10^{-10} rms0.0005 ppm
OCXO (prop.)1 $\times$ 10^{-11} rms0.00001 ppm

Long-Term Stability

The long-term stability of the timebase clock oscillator is due largely to crystal aging. The nature of the crystal, the quality of the crystal, and the plane from which the particular resonator was cut from the original quartz crystal are determining factors in defining aging. This figure is usually given in terms of frequency units per month.

RTXO3 $\times$ 10^{-7}/month0.3 ppm
TXCO1 $\times$ 10^{-7}/month0.1 ppm
OCXO (on/off)1 $\times$ 10^{-7}/month0.1 ppm
OCXO (prop.)1.5 $\times$ 10^{-8}/month0.015 ppm
OXCO (prop.)5 $\times$ 10^{-10}/day0.0005 ppm

EXAMPLE

The long-term stability of a digital frequency counter used to measure the output of a 467.125 MHz transmitter is specified as 0.0015 ppm/mo. What is the error seven months after the last calibration?

$\pm$0.0015 ppm/mo. $\times$ 7 mos. $\times$ 467.125 = $\pm$4.91 Hz

INPUT CIRCUITRY

The purpose of the input circuitry is to condition the signal so that it can be applied to the counter. Radio frequency signals are not usually compatible with the counter circuits, which want to see pulse or square waveforms. Figure 7.7 shows the simplest form of input circuitry. It consists of an attenuator, an amplifier, and a *Schmidt trigger* circuit. In economy counters the amplifier and attenuator might not be present, while in higher-grade instruments they will be present and capable of being switched in and out of the signal path. The purpose of the attenuator and amplifier stages is to adjust the input signal level to the range required to operate the counter.

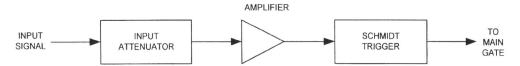

Figure 7.7 Input stages of a digital frequency counter.

Schmidt Trigger

The key to converting the RF sine wave signal into a square wave signal that the counter can accommodate is the Schmidt trigger. Figure 7.8 shows the operation of this circuit. The Schmidt trigger has two limits called the *lower hysteresis limit* (LHL) and the *upper hysteresis limit* (UHL). To produce an output pulse the input signal must slew through both limits in the *right direction*. The Schmidt trigger represented in Figure 7.8 requires the signal to slew through the UHL in a positive-going direction, and then slew through the LHL in a negative-going direction. This action produces a nice clean output pulse that the counter can use from an input that consists of a sine wave signal. When the input signal meets the Schmidt trigger criteria, triggering of the counter can occur.

The voltage difference between the LHL and UHL is called the *trigger window,* while the voltage midway between the LHL and UHL is called the *trigger level.* Some counters allow the setting of either or both of these values from the front panel.

The input stage of Figure 7.8 is a bit simplistic, although it is found on some lower-cost economy counters. The block diagram shown in Figure 7.9 is a bit more

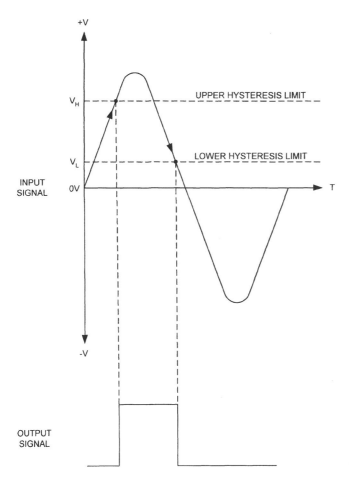

Figure 7.8
Schmidt trigger operation.

representative of counters used in professional service. Like the previous version it contains an attenuator, amplifier, and Schmidt trigger, with the attenuator and amplifier possibly being switchable. There are also several other functions performed in this version.

Input Coupling

Switch S1 allows selection of either AC or DC coupling. When AC coupling is present, capacitor C1 is in series with the signal path, while in DC coupling the path is straight through. AC coupling raises the bottom end frequency range from "near-DC" to about 30 Hz, so it is rarely a factor in RF measurements. The AC coupling position can be used to overcome a DC offset potential present in the signal. In this particular implementation there is also an "input grounded" selection. When this selection is made, the input to the signal conditioning circuit (not the input connector to the counter) is grounded, so the count should go to zero.

Automatic Gain Control (AGC)

The purpose of the AGC is to keep the input signal within a range that can be used by the counter circuit. It will automatically adjust the gain of the circuit by comparing the amplifier output to a desired minimum level.

Fuse and Limiter

Protection against large signals is provided by the fuse and the limiter. The fuse is a form of last resort protection in case anyone does something really stupid, like

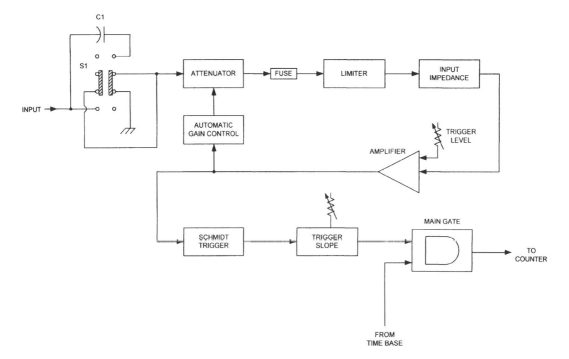

Figure 7.9 Input stages of a practical frequency counter.

connecting the input of the counter directly to the output of a transmitter. The limiter provides protection against less drastic overload.

Input Impedance

A means is also provided to set the input impedance to a specified value. In general, low-frequency counter (<10 MHz) are high-impedance input devices (typically 1-megohm shunted by 20 pF), while above 10 MHz it is 50 ohms.

Trigger Controls

There are two trigger controls: *trigger level* and *trigger slope*. Some counters also include a control that adjusts the width of the trigger window (UHL-LHL).

In some counters the trigger level can be adjusted continuously over a range, while in others it can be set to three different points (as shown in Figure 7.10). The hysteresis window (V_H) remains constant, but the trigger level (midpoint between hysteresis levels) is adjusted. In the three-level version of Figure 7.10 the three levels are *PRESET*, −TRIG, and +TRIG.

The action of the trigger-level function can be seen in Figure 7.11. Keep in mind that the input signal must cross both LHL and UHL or triggering cannot occur. In two of the cases shown the signal does not meet this criterion, so no counting occurs. This can lead to erroneous counting, which will be discussed below.

The trigger slope control is used to determine whether the counter will trigger on the positive slope or negative slope of the input signal waveform.

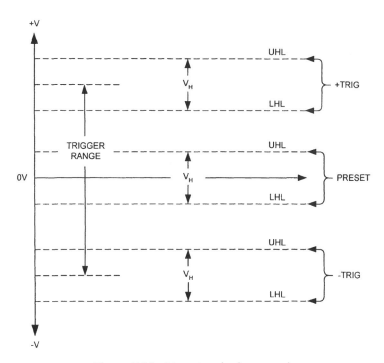

Figure 7.10 Trigger input levels compared.

Counting Errors

There are a number of counter errors, some of which were discussed above relative to the timebase circuits. In this section we will take a look at other forms of error.

Trigger Errors

Figure 7.12A shows the result of misadjusting the trigger-level control on a signal with a heavy amount of harmonic distortion. Note that the sine wave peaks dip markedly as a result of the distortion. When the hysteresis window is set such that the distorted portion of the waveform can cross the UHL and LHL, then multiple counts occur. Figure 7.12B shows how readjusting the trigger-level control can overcome this problem. The distortion excursions are totally outside the range of the hysteresis limits.

The same sort of situation is shown in Figure 7.13A when a ringing pulse is applied to the counter input. If the hysteresis window is misadjusted, a portion of

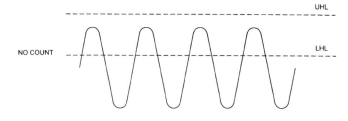

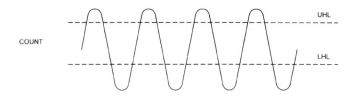

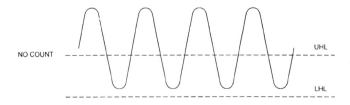

Figure 7.11
Triggering conditions resulting in either counting or not counting input frequency.

the ringing waveform will cross the thresholds, possibly many times. Again, multiple counting occurs. Like the previous case, the problem is solved by adjusting the trigger level to place the ringing waveform outside at least one of the limits so that both cannot be crossed (Figure 7.13B).

±*1-Count Error*

This error occurs because there is no coherence between the input signal and the timebase. In Figure 7.14 the timebase forces the main gate open long enough to allow 10 pulses through. In period T1 all 10 pulses are registered, but in T2 a slight misalignment of the gate opening and the input signal allows only 9 to be registered on the counter.

Because the ±1-count error always involves only one count, the error is inversely proportional to the frequency being measured and the timebase gate period.

$$\pm\ 1\ Count\ Error = \frac{100\%}{f\,t} \tag{7.2}$$

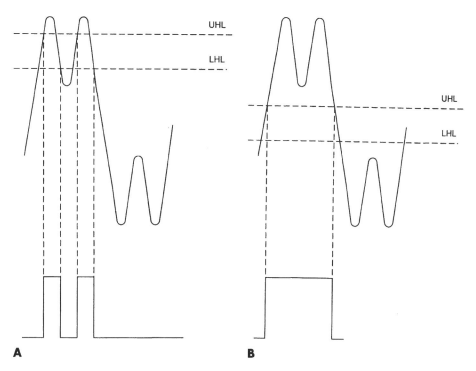

A   **B**

Figure 7.12 (A) Counting error due to complex waveform (e.g., harmonic distortion); (B) Correct threshold setting eliminates the problem.

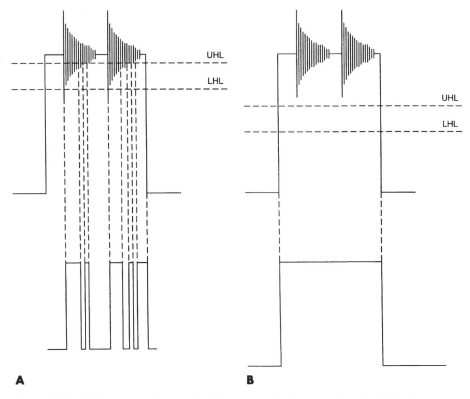

Figure 7.13 (A) Ringing or noise on a signal can cause multiple counting; (B) Threshold setting corrects problem.

Where:

f is the frequency in hertz (Hz)
t is the gate time in seconds (s)

PRESCALED FREQUENCY COUNTERS

Digital frequency counters can be used above their natural frequency range by the simple expedient of dividing the input frequency before applying it to the counter

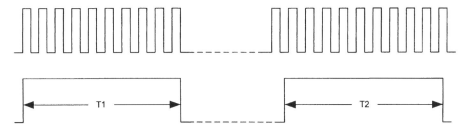

Figure 7.14 One-count error illustrated.

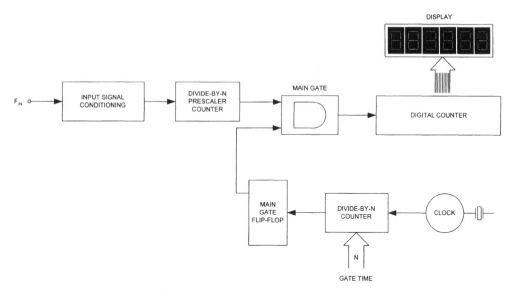

Figure 7.15 Prescaled frequency counter.

input. This is called *prescaling.* Figure 7.15 shows a prescaled frequency counter. Prescaling allows the measurement of frequencies that are higher than the counter could otherwise handle. But there are trade-offs involved. If the input frequency is divided by N, then the gate-opening time must be scaled upwards by the same factor N.

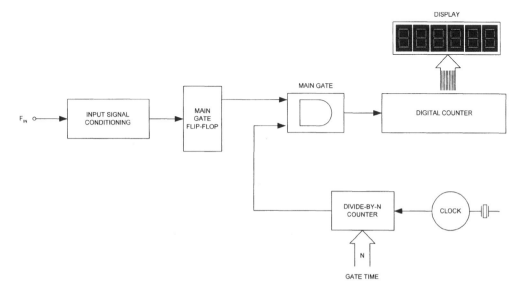

Figure 7.16 Period counter block diagram.

Period Counters

A period counter measures the length of time required for one cycle of the input waveform. Figure 7.16 shows the block diagram for a period counter. Note that it is essentially the same as the frequency counter, but with the roles of the input signal and timebase signal reversed. The main gate flip-flop is controlled by the input signal. It will open the gate for a period of time equal to one cycle of the input signal, allowing timebase pulses to flow into the counter.

The principal use for period counters is measurement of low-frequency signals. Subaudio tones can be measured to a much greater accuracy than with a frequency counter. For example, suppose a 25-Hz signaling tone is measured. If an 8-digit counter with a 1-second gate time is used, then this frequency is displayed at 0.025 kHz. If a period counter is used, however, a much more accurate reading is obtained. The period of a 25-Hz signal is $1/25 = 0.04$ s. If a 1-MHz timebase frequency is used, then the number of pulses counted will be 1,000,000 Hz $\times$ 0.040 s = 40,000 counts = 40,000 μs. This value can be converted to frequency by converting it to seconds and taking the reciprocal. Figure 7.17 shows an example of a portable frequency counter that operates to 3 GHz.

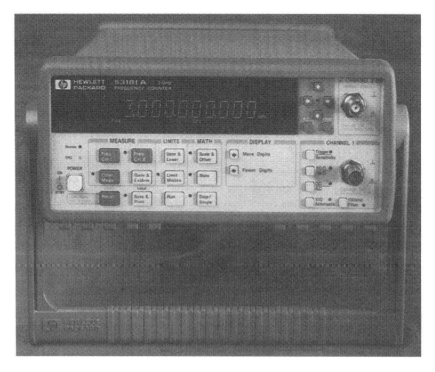

Figure 7.17 Example of a commercial digital frequency counter. (Photo courtesy of Hewlett-Packard)

CHAPTER EIGHT

Radio Receivers and Their Measurements

Radio receivers are at the heart of all communications systems. Wireless telecommunications, broadcasting, and radar could not exist without radio receivers. Navigation depends on radio receivers today. The Global Positioning System (GPS) receiver used in maritime, airborne, and land navigation systems uses a radio receiver at its heart. In this chapter we will cover several topics. We will discuss the different types of radio receivers that are on the market, as well as how to interpret receiver specifications. Finally, we will discuss the testing approaches.

WHAT IS A RADIO RECEIVER?

A radio receiver is an electronic device that must perform two basic functions: (a) *It must respond to, detect, and demodulate desired signals;* and (b) *It must not respond to, detect, or be adversely affected by undesired signals.* If it fails in either of these two functions, then it is a poorly performing design. Both functions are necessary. Weakness in either function makes a receiver a poor bargain, unless there is some mitigating circumstance. The receiver's performance specifications tell us the manufacturer's claims about how well their product performs these two functions. The tests associated with radio receivers are designed to measure either or both of these attributes.

Crystal Video Receivers

Crystal video receivers (Figure 8.1) grew out of primordial crystal sets, but are used in microwave bands even today. The original crystal sets used a naturally occurring PN junction "diode" made from a natural lead compound called galena crystal with an inductor-capacitor (L-C) tuned circuit. Later, crystal sets could be made using germanium or silicon diodes. When vacuum tubes became generally available, it was common to place an audio amplifier at the output of the crystal set. Modern crystal video receivers use silicon or gallium-arsenide microwave diodes

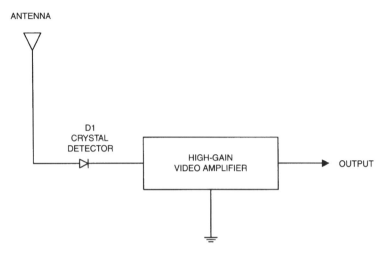

Figure 8.1 Crystal video receiver block diagram.

and a wideband video amplifier (rather than the audio amplifier). Applications include some speed radar receivers, aircraft radar warning receivers, and some very short-range communications receivers.

Tuned Radio Frequency (TRF) Receivers

The tuned radio frequency (TRF) radio receiver uses an L-C resonant circuit in the front-end, followed by one or more radio frequency amplifiers followed by a detector stage. Two varieties are shown in Figures 8.2 and 8.3. The version in Figure 8.2 is called a *tuned gain-block receiver.* It is commonly used in certain VLF scientific applications. Early TRF models (1920s) used independently tuned L-C circuits, but those proved to be very difficult to tune without creating an impromptu Miller oscillator circuit. Later versions mechanically linked ("ganged") the tuned circuits to operate from a single tuning knob, as in Figure 8.3.

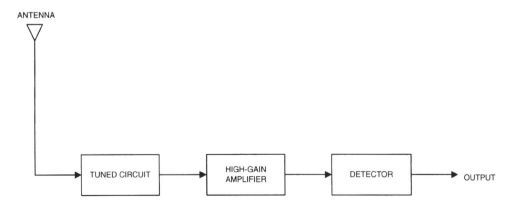

Figure 8.2 Simple-tuned radio frequency receiver block diagram.

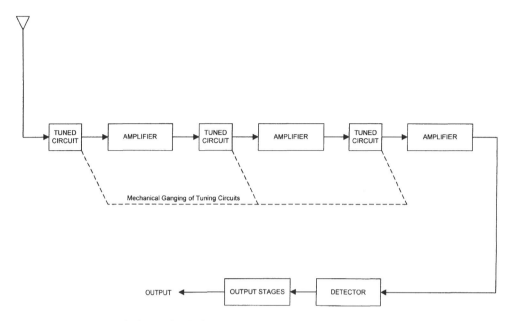

Figure 8.3 Multiply-tuned radio frequency receiver block diagram

Superheterodyne Receivers

Figure 8.4 shows the block diagram of a superheterodyne receiver. We will use this hypothetical receiver as the basic generic framework for evaluating receiver performance and the associated tests. The design in Figure 8.4 is called a *superheterodyne* receiver and represents the largest class of radio receivers in use today.

The superheterodyne receiver block diagram in Figure 8.4 is typical of many receivers. The purpose of a superheterodyne is to convert the incoming RF frequency to a single frequency where most of the signal processing takes place. The front-end section of the receiver consists of the *radio frequency* (RF) amplifier and any RF tuning circuits that may be used (A-B-C in Figure 8.4). In some cases, the RF tuning is very narrow, and basically tunes one frequency. In other cases, the RF front-end tuning is broadbanded. In that case, bandpass filters are used.

The frequency translator section (D and E) is also considered part of the front-end in most textbooks, but here we will label it as a separate entity. The translator consists of a frequency mixer and a local oscillator. This section does the heterodyning, which is discussed in more detail below. The output of the frequency translator is called the *intermediate frequency* (IF).

The translator stage is followed by the intermediate frequency (IF) amplifier. The IF amplifier (F-G-H) is basically a radio frequency amplifier tuned to a single frequency. The IF can be higher or lower than the RF frequency, but it will always be a single frequency.

A sample of the IF amplifier output signal is applied to an *automatic gain control* (AGC) section (L-M). The purpose of this section is to keep the signal level in the output more or less constant. The AGC circuit consists of a rectifier and ripple

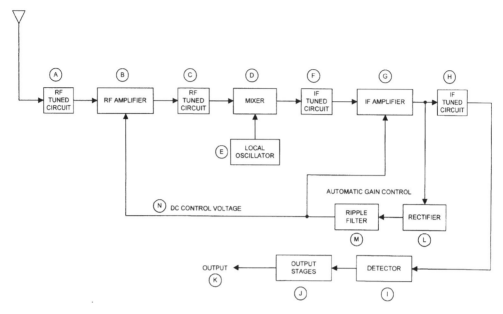

Figure 8.4 Superheterodyne receiver block diagram.

filter that produces a DC control voltage. The DC control voltage is proportional to the input RF signal level (N). It is applied to the IF and RF amplifiers to raise or lower the gain according to signal level. If the signal is weak, then the gain is forced higher, and if the signal is strong, the gain is lowered. The end result is to smooth out variations of the output level.

The detector stage (I) is used to recover any modulation on the input RF signal. The type of detector depends on the type of modulation used for the incoming signal. Amplitude modulation (AM) signals are generally handled in an *envelope detector.* In radio astronomy a special variant of the envelope detector called a *square law detector* is used. The difference is that the straight envelope detector is linear, while the square law detector is nonlinear to voltage. The square law detector takes advantage of the fact that the power level of the signal is related to the square of the applied signal voltage.

The output stages (J-K) are used to amplify and deliver the recovered modulation to the user. If the receiver is for broadcast use, then the output stages are audio amplifiers and loudspeakers. In some scientific receivers the output stages consist of integrator circuits and (sometimes) DC amplifiers.

Heterodyning

The main attribute of the superheterodyne receiver is that it converts the radio signal's RF frequency to a standard frequency for further processing. Although today the new frequency, called the *intermediate frequency* or *IF,* may be either higher or lower than the RF frequencies, early superheterodyne receivers always downconverted RF signal to a lower IF frequency (IF < RF). The reason was purely practical, for in those days higher frequencies were more difficult to process than lower frequencies. Even today, because variable-tuned circuits still tend to offer different performance over the band being tuned, converting to a single IF frequency, and obtaining most of the gain and selectivity functions at the IF, allows a more uniform overall performance over the entire range being tuned.

A superheterodyne receiver works by frequency converting (or simply "heterodyning"—the added "super" is vintage 1920s advertising hype) the RF signal. This occurs by nonlinearly mixing the incoming RF signal with a *local oscillator* (LO) signal. When this process is done, disregarding noise, the output spectrum will contain a large variety of signals according to:

$$F_O = mF_{RF} \pm nF_{LO} \qquad\qquad [8.1]$$

Where:

F_{RF} is the frequency of the RF signal

F_{LO} is the frequency of the local oscillator

m and n are either zero or integers (0, 1, 2, 3 . . . n)

Equation 8.1 means that there will be a large number of signals at the output of the mixer, although for the most part the only ones that are of immediate concern to understanding superheterodyne operation are those for which m and n are either 0 or 1. Thus, for our present purpose, the output of the mixer will be the fundamentals (F_{RF} and F_{LO}) and the second-order products ($F_{LO} - F_{RF}$ and $F_{LO} + F_{RF}$), as seen in Figure 8.5. Some mixers, notably those described as *double-balanced mixers* (DBM), suppress F_{RF} and F_{LO} in the mixer output, so only the second-order sum and difference frequencies exist with any appreciable amplitude. This case is simplistic and is used only for this present discussion. Later on, we will look at what happens when third-order ($2F1 \pm F2$ and $2F2 \pm F1$) and fifth-order ($3F1 \pm 2F2$ and $3F2 \pm 2F1$) become large.

Note that the local oscillator frequency can be either higher than the RF frequency (*high-side injection*) or lower than the RF frequency (*low-side injection*). There is ordinarily no practical reason to prefer one over the other except that it will make a difference whether the main tuning dial reads high-to-low or low-to-high.

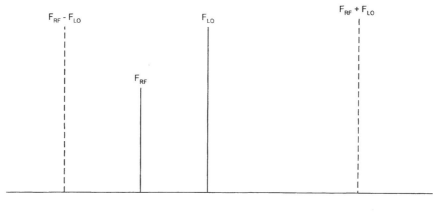

Figure 8.5 Simplified heterodyning mixer spectrum uses either sum or difference between LO and RF signals to produce an Intermediate Frequency (IF).

The candidates for IF include the sum (LO + RF) and difference (LO − RF) second-order products found at the output of the mixer. A high-Q tuned circuit following the mixer will select which of the two are used. Consider an example. Suppose an AM broadcast band superheterodyne radio has an IF frequency of 455 KHz, and the tuning range is 540 to 1,700 kHz. Because the IF is lower than any frequency within the tuning range, it will be the difference frequency that is selected for the IF. The local oscillator is set to be high-side injection, so it will tune from (540 + 455) = 995 kHz to (1,700 + 455) = 2,155 kHz.

Front-End Circuits

The principal task of the front-end and frequency translator sections of the receiver in Figure 8.4 is to select the signal and convert it to the IF frequency. But in many radio receivers there may be additional functions. In some cases (but not all), an RF amplifier will be used in front of the mixer. Typically, these amplifiers have a gain of 3 to 10 dB, with 5 to 6 dB being very common. The tuning for the RF amplifier is sometimes a broad bandpass-fixed frequency filter that admits an entire band. In other cases, it is a narrow-band, but variable-frequency, tuned circuit.

Intermediate Frequency (IF) Amplifier

The IF amplifier is responsible for providing most of the gain in the receiver, as well as the narrowest bandpass filtering. It is a high gain, usually multistaged, single-frequency tuned radio frequency amplifier. For example, one HF shortwave receiver block diagram lists 120 dB of gain from antenna terminals to audio output, of which 85 dB are provided in the 8.83 MHz IF amplifier chain. In the example of Figure 8.4, the receiver is a *single conversion* design, so there is only one IF amplifier section.

Detector

The detector demodulates the RF signal and recovers whatever audio (or other information) that will be heard by the listener. In a straight AM receiver, the detector will be an ordinary half-wave rectifier and ripple filter; it is called an *envelope detector*. In other detectors—notably double sideband suppressed carrier (DSBSC), single sideband suppressed carrier (SSBSC or SSB), or continuous wave (CW or morse telegraphy)—a second local oscillator, usually called a *beat frequency oscillator* (BFO), which is operating near the IF frequency, is heterodyned with the IF signal. The resultant difference signal is the recovered audio. That type of detector is called a *product detector*. Many AM receivers today have a sophisticated *synchronous detector*, rather than the simple envelope detector. Receivers that accept frequency modulation (FM) will provide some sort of frequency or phase-sensitive detector such as a ratio detector, discriminator, quadrature detector, phase-locked loop detector, and so forth.

Audio Amplifiers

The audio amplifiers are used to finish the signal processing. They also boost the output of the detector to a usable level to drive a loudspeaker or set of earphones. The audio amplifiers are sometimes used to provide additional filtering. It is quite common to find narrow-band filters to restrict audio bandwidth, or notch filters to eliminate interfering signals that make it through the IF amplifiers intact.

There are three basic areas of receiver performance that must be considered. Although interrelated, they are sufficiently different to merit individual consideration: *noise, static,* and *dynamic.* We will look at all of these areas, but first let's look at the units of measure that we will use in this series.

Units of Measure

Input Signal Voltage

Input signal level, when specified as a voltage, is typically stated in either *microvolts* (μV) or *nanovolts* (nV). The volt is simply too large a unit for practical use on radio receivers. Signal input voltage (or sometimes power level) is often used as part of the *sensitivity* specification, or as a test condition for measuring certain other performance parameters.

There are two forms of signal voltage that are used for input voltage specification: *source voltage* (V_{EMF}) and *potential difference* (V_{PD}), as illustrated in Figure 8.6. The source voltage (V_{EMF}) is the open terminal (no load) voltage of the signal generator or source, while the potential difference (V_{PD}) is the voltage that appears across the receiver antenna terminals with the load connected (the load is the receiver antenna input impedance, R_{in}). When $R_s = R_{in}$, the preferred "matched impedances" case in radio receiver systems, the value of V_{PD} is one-half V_{EMF}. This can be seen in Figure 8.6 by noting that R_S and R_{in} form a voltage divider network driven by V_{EMF}, with V_{PD} as the output.

dBm

These units refer to *decibels relative to one milliwatt (1 mW) dissipated in a 50-ohm resistive impedance* (defined as the 0-dBm reference level) and is calculated from:

$$dBm = 10 \, LOG \left[\frac{P_{Watts}}{0.001} \right] \qquad [8.2]$$

or,

$$dBm = 10 \, LOG \, (P_{MW}) \qquad [8.3]$$

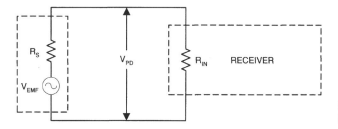

Figure 8.6
Input signal level definitions.

In the noise voltage case calculated above, 0.028 μV in 50 ohms, the power is $V^2/50$, or 5.6×10^{-10} watts, which is 5.6×10^{-7} mw. In dBm notation, this value is 10 LOG (5.6×10^{-7}), or -62.5 dBm.

dBmV

This unit is used in television receiver systems in which the system impedance is 75 ohms, rather than the 50 ohms normally used in other RF systems. It refers to the signal voltage, measured in decibels, with respect to a signal level of one millivolt (1 mV) across a 75-ohm resistance (0 dBmv). In many TV specs, 1 mV is the full quieting signal that produces no "snow" (i.e., noise) in the displayed picture. Note: 1 mV = 1,000 μV.

dBμV

This unit refers to a signal voltage, measured in decibels, relative to one microvolt (1 μV) developed across a 50-ohm resistive impedance (0 dBμV). For the case of our noise signal voltage, the level is 0.028 μV, which is the same as -31.1 dBμV. The voltage used for this measurement is usually the V_{EMF}, so to find V_{PD} divide it by two after converting dBμV to μV. To convert dBμV to dBm, merely subtract 113; that is, 100 dBμV = -13 dBm.

It requires only a little algebra to convert signal levels from one unit of measure to another. This job is sometimes necessary when a receiver manufacturer mixes methods in the same specifications sheet. In the case of dBm and dBμV, 0 dBμV is 1 μV V_{EMF}, or a V_{PD} of 0.5 μV, applied across 50 ohms, so the power dissipated is 5×10^{-15} watts, or -113 dBm.

NOISE

A radio receiver must detect signals in the presence of noise. The *signal-to-noise ratio* (SNR) is the key here, because a signal must be above the noise level before it can be successfully detected and used.

Noise comes from a number of different sources, but for the sake of this discussion we can divide them into two classes: *sources external to the receiver* and *sources internal to the receiver*. There is little one can do about the external noise sources, for they consist of natural and manmade electromagnetic signals that fall within the passband of the receiver. Figure 8.7 shows an approximation of the external noise situation from the middle of the AM broadcast band to the low end of the VHF region. One must select a receiver that can cope with external noise sources, especially if the noise sources are strong.

Some natural external noise sources are extraterrestrial. It is these signals that form the basis of radio astronomy. For example, if you aim a beam antenna at the eastern horizon prior to sunrise, a distinct rise of noise level occurs as the sun slips above the horizon, especially in the VHF region. The reverse occurs in the West at sunset, but is less dramatic, probably because atmospheric ionization decays much slower than it is generated. During World War II, it is reported that British radar operators noted an increase in received noise level any time the Milky Way was above the horizon, decreasing the range at which they could detect in-bound

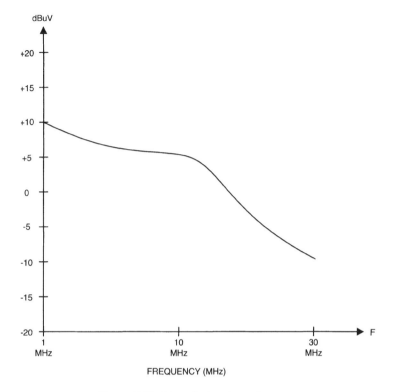

Figure 8.7 Atmospheric noise vs. frequency.

bombers. There is also some well-known, easily observed noise from the planet Jupiter in the 18- to 30-MHz band.

The receiver's internal noise sources are affected by the design of the receiver. Ideal receivers produce no noise of their own, so the output signal from the ideal receiver would contain only the noise that was present at the input along with the radio signal. But real receiver circuits produce a certain level of internal noise of their own. Even a simple fixed-value resistor is noisy. Figure 8.8A shows the equivalent circuit for an ideal, noise-free resistor, while Figure 8.8B shows a practical real-world resistor. The noise in the real-world resistor is represented in Figure 8.8B by a noise voltage source, V_n, in series with the ideal, noise-free resistance, R_i. At any temperature above *Absolute Zero* (0 °K, or about −273 °C) electrons in any material are in constant random motion. Because of the inherent randomness of that motion, however, there is no detectable current in any one direction. In other words, electron drift in any single direction is canceled over even short time periods by equal drift in the opposite direction. Electron motions are therefore statistically decorrelated. There is, however, a continuous series of random current pulses generated in the material, and those pulses are seen by the outside world as noise signals.

If a shielded 50-ohm resistor is connected across the antenna input terminals of a radio receiver, the noise level at the receiver output will increase by a predictable amount over the short-circuit noise level. Noise signals of this type are

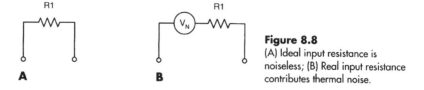

Figure 8.8
(A) Ideal input resistance is noiseless; (B) Real input resistance contributes thermal noise.

called by several names: *thermal agitation noise, thermal noise,* or *Johnson noise*. This type of noise is also called "white noise" because it has a very broadband (nearly gaussian) spectral density. The thermal noise spectrum is dominated by midfrequencies (10^4 to 10^5 Hz) and is essentially flat. The term "white noise" is a metaphor borrowed from white light, which is composed of all visible color frequencies. The expression for such noise is:

$$V_N = \sqrt{4\,K\,T\,B\,R} \qquad\qquad [8.4]$$

Where:

V_n is the noise potential in volts (V)

K is Boltzmann's constant (1.38×10^{-23} J/°K)

T is the temperature in degrees Kelvin (°K), normally set to 290 or 300 °K by convention.

R is the resistance in ohms (Ω)

B is the bandwidth in hertz (Hz)

Table 8.1 and Figure 8.9 show noise values for a 50-ohm resistor at various bandwidths out to 5 and 10 kHz, respectively. Because different bandwidths are used for different reception modes, it is common practice to delete the bandwidth factor in Equation 8.4 and to write it in the form:

$$V_N = \sqrt{4\,K\,T\,R} \quad V/\sqrt{Hz} \qquad\qquad [8.5]$$

With Equation 8.5 one can find the noise voltage for any particular bandwidth by taking its square root and multiplying it by the equation. This equation is essentially the solution of the previous equation normalized for a 1-Hz bandwidth.

Signal-to-Noise Ratio (SNR or S_n)

Receivers are evaluated for quality on the basis of *signal-to-noise ratio* (SNR or S/N), sometimes denoted S_n. The goal of the designer is to enhance the SNR as much as possible. Ultimately, the minimum signal level detectable at the output of an amplifier or radio receiver is that level that appears just above the noise floor

TABLE 8.1 Bandwidth and thermal noise.

Bandwith (Hz)	Noise Voltage
1000	2.83E-08
1500	3.46E-08
2000	4.00E-08
2500	4.47E-08
3000	4.90E-08
3500	5.29E-08
4000	5.66E-08
4500	6.00E-08
5000	6.33E-08
5500	6.63E-08
6000	6.93E-08
6500	7.21E-08
7000	7.49E-08
7500	7.75E-08
8000	8.00E-08
8500	8.25E-08
9000	8.49E-08
9500	8.72E-08
10000	8.95E-08

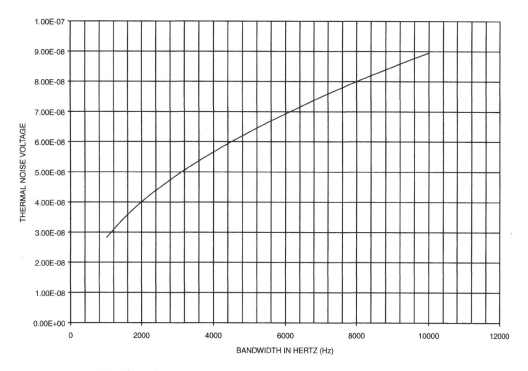

Figure 8.9 Thermal noise vs. receiver bandwidth.

level. Therefore, the lower the system noise floor, the smaller the *minimum allowable signal*.

Noise Factor, Noise Figure, and Noise Temperature

The noise performance of a receiver or amplifier can be defined in three different but related ways: *noise factor* (F_n), *noise figure* (NF), and *equivalent noise temperature* (T_e); these properties are definable as a simple ratio, decibel ratio, or Kelvin temperature, respectively.

Noise Factor (F_n)

For components such as resistors, the noise factor is the ratio of the noise produced by a real resistor to the simple thermal noise of an ideal resistor.

The noise factor of a radio receiver (or any system) is the ratio of output noise power (P_{no}) to input noise power (P_{ni}):

$$F_N = \left[\frac{P_{NO}}{P_{NI}} \right]_{T=290\ ^\circ K} \qquad [8.6]$$

In order to make comparisons easier, the noise factor is usually measured at the standard temperature (T_o) of 290 °K (standardized room temperature). In some countries, however, 299 °K or 300 °K are commonly used (the differences are negligible). It is also possible to define noise factor F_N in terms of the output and input signal-to-noise ratios:

$$F_N = \frac{S_{NI}}{S_{NO}} \qquad [8.7]$$

Where:

S_{NI} is the input signal-to-noise ratio
S_{NO} is the output signal-to-noise ratio

Noise Figure (NF)

The *noise figure* is frequently used to measure the receiver's "goodness," that is, its departure from "idealness." Thus, it is a *figure of merit*. The noise figure is the noise factor converted to decibel notation:

$$\text{N.F.} = 10\ \text{LOG}\ (F_N) \qquad [8.8]$$

Where:

N.F. is the noise figure in decibels (dB)
F_n is the noise factor
LOG refers to the system of base-10 logarithms

Noise Temperature (T$_e$)

The noise "temperature" is a means for specifying noise in terms of an equivalent temperature. That is, the noise level that would be produced by a resistor at that temperature (expressed in degrees Kelvin). Evaluating the noise equations shows that the noise power is directly proportional to temperature in degrees Kelvin, and also that noise power collapses to zero at the temperature of Absolute Zero (0 °K).

Note that the equivalent noise temperature T_e is *not* the physical temperature of the amplifier, but rather a theoretical construct that is an *equivalent* temperature that produces that amount of noise power in a resistor. The noise temperature is related to the noise factor by:

$$T_e = (F_N - 1)T_o \qquad [8.9]$$

and to noise figure by

$$T_e = 290 \left[10^{(N.F./10)} - 1\right] \qquad [8.10]$$

Noise temperature is often specified for receivers and amplifiers in combination with, or in lieu of, the noise figure.

Noise in Cascade Amplifiers

A noise signal is seen by any amplifier following the noise source as a valid input signal. Each stage in the cascade chain (Figure 8.10) amplifies both signals and noise from previous stages and also contributes some additional noise of its own. Thus, in a cascade amplifier, the final stage sees an input signal that consists of the original signal and noise amplified by each successive stage, plus the noise contributed by earlier stages. The overall noise factor for a cascade amplifier can be calculated from *Friis' noise equation:*

$$F_N = F_1 + \frac{F_2 - 1}{G1} + \frac{F_3 - 1}{G1\,G2} + \cdots + \frac{F_N - 1}{G1\,G2\,\ldots\,G_{N-1}} \qquad [8.11]$$

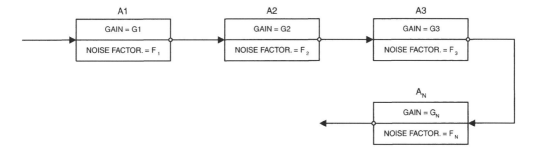

Figure 8.10 Three-stage amplifier for gain/noise figure calculations.

Where:

F_n is the overall noise factor of N stages in cascade
F_1 is the noise factor of stage-1
F_2 is the noise factor of stage-2
F_n is the noise factor of the nth stage
G1 is the gain of stage-1
G2 is the gain of stage-2
G_{n-1} is the gain of stage $(n-1)$

As you can see from Friis' equation, the noise factor of the entire cascade chain is dominated by the noise contribution of the first stage or two. High-gain, low-noise radio astronomy RF amplifiers typically use low-noise amplifier (LNA) circuits for the first stage or two in the cascade chain. Thus, you will find an LNA at the feedpoint of a satellite receiver's dish antenna, and possibly another one at the input of the receiver module itself, but other amplifiers in the chain might be more modest (although their noise contribution cannot be ignored at radio astronomy signal levels).

The matter of signal-to-noise ratio (SNR) is sometimes treated in different ways that each attempt to crank some reality into the process. The signal-plus-noise-to-noise ratio $(S+N/N)$ is found quite often. As the ratios get higher, the SNR and $S+N/N$ converge (only about 0.5 dB difference at ratios as little as 10 dB). Still another variant is the SINAD (signal-plus-noise-plus-distortion-to-noise) ratio. The SINAD measurement takes into account most of the factors that can deteriorate reception.

Receiver Noise Floor

The *noise floor* of the receiver is a statement of the amount of noise produced by the receiver's internal circuitry and directly affects the *sensitivity* of the receiver. The noise floor is typically expressed in dBm. The noise floor specification is evaluated as follows: the more negative the better. The best receivers have noise floor numbers of less than −130 dBm, while some very good receivers offer numbers of −115 to −130 dBm.

The noise floor is directly dependent on the bandwidth used to make the measurement. Receiver advertisements usually specify the bandwidth, but note whether or not the bandwidth that produced the very good performance numbers is also the bandwidth that you will need for the mode of transmission you want to receive. If, for example, you are interested only in weak 6-kHz-wide AM signals, and the noise floor is specified for a 250 Hz CW filter, then the noise floor might be too high for your use.

Static Measures of Receiver Performance

The two principal static levels of performance for radio receivers are *sensitivity* and *selectivity*. The sensitivity refers to the level of input signal required to produce a usable output signal (variously defined). The selectivity refers to the ability of the receiver to reject adjacent channel signals (again, variously defined). Let's take a look at both of these factors. Keep in mind, however, that in modern high-performance radio receivers the static measures of performance may also be the least relevant compared with the dynamic measures.

Sensitivity

Sensitivity is a measure of the receiver's ability to pick up ("detect") signals, and is usually specified in microvolts (μV). A typical specification might be "0.5 μV sensitivity." The question to ask is: "relative to what?" The sensitivity number in microvolts is meaningless unless the test conditions are specified. For most commercial receivers, the usual test condition is the sensitivity required to produce a 10 dB signal-plus-noise-to-noise (S + N/N) ratio in the mode of interest. For example, if only one sensitivity figure is given, one must find out what bandwidth is being used. The normal bandwidth is 5 to 6 kHz for AM, 2.6 to 3 kHz for single sideband, 1.8 kHz for radioteletype, or 200 to 500 Hz for CW. Radio astronomy receiver bandwidths tend to be much wider, so sensitivity figures must be adjusted.

The amount of sensitivity improvement is seen by evaluating some simple numbers. Recall that a claim of "x $-\mu$V" sensitivity refers to some standard such as "x $-\mu$V to produce a 10 dB signal-to-noise ratio in y-Hz bandwidth." Consider the case where the main mode for a high frequency (HF) shortwave receiver is AM (for international broadcasting), the sensitivity is 1.9 μV for 10 dB SNR, and the bandwidth is 5 kHz. If the bandwidth were reduced to 2.8 kHz for SSB, then the sensitivity improves by the square root of the ratio, or $\sqrt{5}/2.8$. If the bandwidth is further reduced to 270 Hz (i.e., 0.27 kHz) for CW, then the sensitivity for 10 dB SNR is $\sqrt{5}/0.27$. The 1.9 μV AM sensitivity therefore translates to 1.42 μV for SSB and 0.44 μV for CW. If only the CW version is given, then the receiver might be made to look a whole lot better than it is, even though the typical user may never use the CW mode (see Figure 8.11).

The sensitivity differences also explain why weak SSB signals can be heard under conditions when AM signals of similar strength have disappeared into the noise, or why the CW mode has as much as a 20-dB advantage over SSB, *ceteris paribus*.

In some receivers, the difference in mode (AM, SSB, RTTY, CW, etc.) can conceivably result in sensitivity differences that are more than the differences in the bandwidths associated with the various modes. The reason is that there is sometimes a "processing gain" associated with the type of detector circuit used to demodulate the signal at the output of the IF amplifier. A simple AM envelope detector is lossy because it consists of a simple diode (1N60, 1N34, etc.) and an R-C filter (a passive circuit). Other detectors (e.g., product detector for SSB, or synchronous AM detectors) have their own signal gain, so they may produce better sensitivity numbers than the bandwidth suggests.

Another indication of sensitivity is *minimum detectable signal* (MDS), which is usually specified in dBm. This signal level is the signal power at the antenna input terminal of the receiver required to produce some standard S + N/N ratio, such as 3 dB (Figure 8.12). The MDS is actually misnamed, for there is good evidence that trained, experienced radio operators can filter out noise in their ears (or brains), and detect signals buried as much as 10 dB below the MDS. But such skills are not

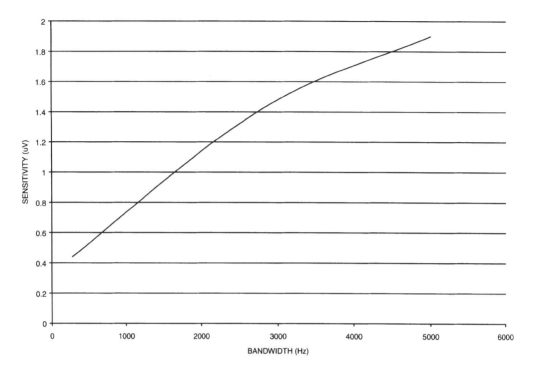

Figure 8.11 Sensitivity vs. bandwidth.

the norm. The MDS is an example of an *operational definition,* that is, a procedure that yields coherent and usable results: MDS is 3 dB above the noise floor.

In radar receivers, the MDS is usually described in terms of a single-pulse return and a specified S + N/N ratio. Also, in radar and other pulse receivers, the sensitivity can be improved by integrating multiple pulses. If N return pulses are integrated, then the sensitivity is improved by a factor of N if coherent detection is used, and $\sqrt{N}$ if noncoherent detection is used.

The noise floor cannot ever be zero because of thermal agitation in resistances and impedances. If the receiver antenna input impedance is 50 ohms, then there will be a −174 dBm/Hz thermal noise level present. As bandwidth increases, then the noise increases proportionally. For example, in an SSB receiver that has a 3,000 Hz IF bandwidth, the noise level will be 10 LOG (3,000) = 34.8 dB higher, or (−174 dBm) + 34.8 dB = −139 dBm.

Modulated signals represent a special case. For those sensitivities, it is common to specify the conditions under which the measurement is made. For example, in AM receivers the sensitivity to achieve 10 dB SNR is measured with the input signal modulated 30% by a 400 or 1,000 Hz sinusoidal tone.

An alternate method is sometimes used for AM sensitivity measurements, especially in servicing consumer radio receivers (where SNR may be a little hard to measure with the equipment normally available to technicians who work on those radios). This is the "standard output conditions" method. Some manuals will specify the audio signal power or audio signal voltage at some critical point, when the

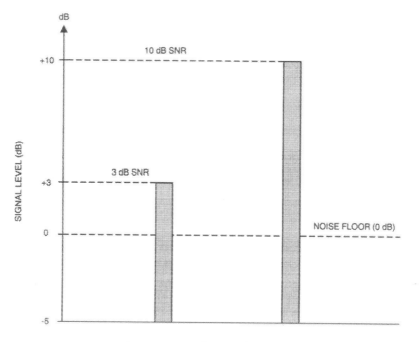

Figure 8.12 Definitions of sensitivity.

30% modulated RF carrier is present. In one automobile radio receiver, the sensitivity was specified as "X μV to produce 400 mW across 8 ohm resistive load substituted for the loudspeaker when the signal generator is modulated 30% with a 400 Hz audio tone." The cryptic note on the schematic showed an output sine wave across the loudspeaker with the label "400 mW in 8Ω (1.79 volts), @30% mod. 400 Hz, 1 μV RF."

The sensitivity is sometimes measured essentially the same way, but the signal levels will specify the voltage level that will appear at the top of the volume control, or output of the detector-filter, when the standard signal is applied. Thus, there are two ways seen for specifying AM sensitivity: *10 dB SNR* and *standard output conditions*.

There are also two ways to specify FM receiver sensitivity. The first is the 10 dB SNR method discussed above, that is, the number of microvolts of signal at the input terminals required to produce a 10 dB SNR when the carrier is modulated by a standard amount. The measure of FM modulation is *deviation* expressed in kilohertz. Sometimes, the full deviation for that class of receiver is used, while for others a value that is 25 to 35% of full deviation is specified.

The second way to measure FM sensitivity is the level of signal required to reduce the no-signal noise level by 20 dB. This is the "20-dB quieting sensitivity of the receiver." If you tune between signals on an FM receiver, you will hear a loud "hiss" signal, especially in the VHF/UHF bands. Some of that noise is externally generated, while some is internally generated. When an FM signal appears in the passband, that hiss is suppressed, even if the FM carrier is unmodulated. The *quieting sensitivity* of an FM receiver is a statement of the number of microvolts required to produce some standard quieting level, usually 20 dB.

Pulse receivers, such as radar and pulse communications units, often use the *tangential sensitivity* as the measure of performance, which is the amplitude of pulse signal required to raise the noise level by its own RMS amplitude (Figure 8.13).

Selectivity

Although no receiver specification is unimportant, if one had to choose between sensitivity and selectivity, the proper choice most of the time would be to take selectivity.

Selectivity is the measure of a receiver's ability to reject adjacent channel interference. Or put another way, it's the ability to reject interference from signals on frequencies close to the desired signal frequency.

In order to understand selectivity requirements, one must first understand a little bit of the nature of radio signals. An unmodulated radio carrier theoretically has an infinitesimal (near-zero) bandwidth (although all real unmodulated carriers have a very narrow, but nonzero, bandwidth because they are modulated by noise and other artifacts). As soon as the radio signal is modulated to carry information, however, the bandwidth spreads. Even an on/off telegraphy (CW) or pulse signal spreads on either side of the carrier frequency an amount that is dependent on the sending speed and the shape of the keying waveform.

An AM signal spreads out an amount equal to twice the highest audio modulating frequencies. For example, a communications AM transmitter will have audio components from 300 to 3,000 Hz, so the AM waveform will occupy a spectrum that is equal to the carrier frequency (F) plus or minus the audio bandwidth (F $\pm$ 3,000 Hz in the case cited). An FM carrier spreads out according to the *deviation*. For example, a narrow-band FM landmobile transmitter with 5 kHz deviation spreads out $\pm$5 kHz, while FM broadcast transmitters spread out $\pm$75 kHz

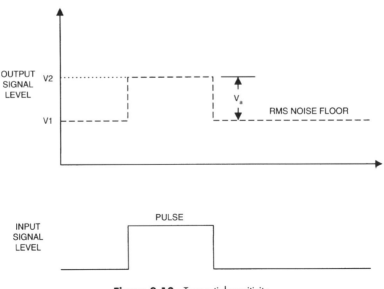

Figure 8.13 Tangential sensitivity.

An implication of the fact that radio signals have bandwidth is that the receiver must have sufficient bandwidth to recover the entire signal. Otherwise, information may be lost and the output is distorted. On the other hand, allowing too much bandwidth increases the noise picked up by the receiver and thereby deteriorates the SNR. The goal of the selectivity system of the receiver is to match the bandwidth of the receiver to that of the signal. That is why receivers will use 270 or 500 Hz bandwidth for CW, 2 to 3 kHz for SSB, and 4 to 6 kHz for AM signals. They allow you to match the receiver bandwidth to the transmission type.

The selectivity of a receiver has a number of aspects that must be considered: *front-end bandwidth, IF bandwidth, IF shape factor,* and the *ultimate* (distant frequency) *rejection.*

Front-End Bandwidth

The "front-end" of a modern superheterodyne radio receiver is the circuitry between the antenna input terminal and the output of the first mixer stage. The reason why front-end selectivity is important is to keep out-of-band signals from afflicting the receiver. For example, AM broadcast band transmitters located nearby can easily overload a poorly designed shortwave or VLF/LF receiver. Even if these signals are not heard by the operator (as they often are), they can desensitize a receiver, or create harmonics and intermodulation products that show up as "birdies" or other types of interference on the receiver. Strong local signals can take a lot of the receiver's dynamic range and thereby make it harder to hear weak signals.

In some "crystal video" microwave receivers, that front-end might be wide open without any selectivity at all, but in nearly all other receivers there will be some form of frequency selection present.

Two forms of frequency selection are typically found. A designer may choose to use only one of them in a design. Alternatively, both might be used in the design, but separately (operator selection). Or finally, both might be used together. These forms can be called the *resonant frequency filter* (Figure 8.14A) and *bandpass filter* (Figure 8.14B) approaches.

The resonant frequency approach uses L-C elements tuned to the desired frequency to select which RF signals reach the mixer. In some receivers, these L-C elements are designed to track with the local oscillator that sets the operating frequency. That is why you see two-section variable capacitors for AM broadcast receivers with two different capacitance ranges for the two sections. One section tunes the LO and the other section tunes the tracking RF input. In other designs, a separate tuning knob ("preselector" or "antenna") is used.

The other approach uses a suboctave bandpass filter to admit only a portion of the RF spectrum into the front-end. For example, a shortwave receiver that is designed to take the HF spectrum in 1-MHz pieces may have an array of RF input bandpass filters that are each 1 MHz wide (e.g., 9 to 10 MHz).

In addition to the reasons cited above, front-end selectivity also helps improve a receiver's *image rejection* and *1st IF Rejection* capabilities.

Image Rejection

An *image* in a superheterodyne receiver is a signal that appears at twice the IF distant from the desired RF signal; it is also located on the opposite side of the LO frequency from the desired RF signal. In Figure 8.15, a superheterodyne operates with

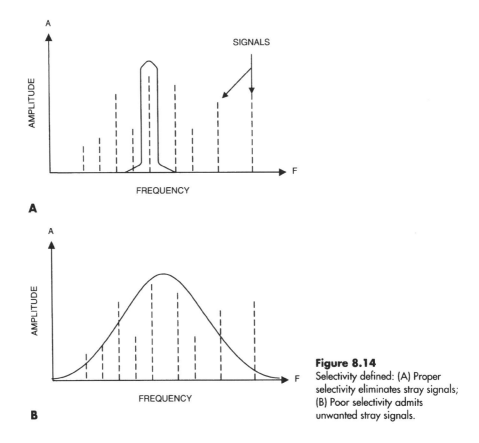

Figure 8.14
Selectivity defined: (A) Proper selectivity eliminates stray signals; (B) Poor selectivity admits unwanted stray signals.

a 455 kHz (i.e., 0.455 MHz) IF, and is turned to 24.0 MHz (F_{RF}). Because this receiver uses low-side LO injection, the LO frequency F_{LO} is 24.0 − 0.455, or 23.545 MHz. If a signal appears at twice the IF below the RF (i.e., 910 kHz below F_{RF}), and reaches the mixer, then it too has a difference frequency of 455 kHz, so it will pass right through the IF filtering as a valid signal. The image rejection specification tells how well this image frequency is suppressed. Normally, anything over about 70 dB is considered good.

Tactics to reduce image response vary with the design of the receiver. The best approach, at design time, is to select an IF frequency that is high enough that the image frequency will fall outside the passband of the receiver front-end. Some modern HF receivers use an IF of 8.83 MHz, 9 MHz, 10.7 MHz, or something similar, and for image rejection these frequencies are considerably better than 455 kHz receivers in the higher HF bands. However, a common trend is to do *double conversion*. In most such designs, the first IF frequency is considerably higher than the RF, being in the range 35 to 60 MHz (50 MHz is common in HF receivers, 70 MHz in microwave receivers).

The high IF makes it possible to suppress the VHF images with a simple low-pass filter. If the 24.0 MHz signal (above) were first up, converted to 50 MHz (74 MHz LO), for example, the image would be at 124 MHz. The second conversion brings the IF down to one of the frequencies mentioned above, or even

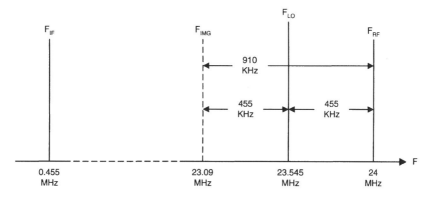

Figure 8.15 Image frequency defined.

455 kHz. The lower frequencies are preferable to 50 MHz for bandwidth selectivity reasons because good quality crystal, ceramic, or mechanical filters in those ranges filters are easily available.

First IF Rejection

The first IF rejection specification refers to how well a receiver rejects radio signals operating on the receiver's first IF frequency. For example, if your receiver has a first IF of 70 MHz, it must be able to reject radio signals operating on that frequency when the receiver is tuned to a different frequency. Although the shielding of the receiver is also an issue with respect to this performance, the front-end selectivity affects how well the receiver performs against first IF signals.

If there is no front-end selectivity to discriminate against signals at the IF frequency, then they arrive at the input of the mixer unimpeded. Depending on the design of the mixer, they then may pass directly through to the high-gain IF amplifiers and be heard in the receiver output.

IF Bandwidth

Most of the selectivity of the receiver is provided by the filtering in the IF amplifier section. The filtering might be L-C filters (especially if the principal IF is a low frequency like 50 kHz), a ceramic resonator, a crystal filter, or a mechanical filter. Of these, the mechanical filter is usually regarded as best for narrow bandwidths, with the crystal filter and ceramic filter coming in next.

The IF bandwidth is expressed in kilohertz (kHz), and is measured from the points on the IF frequency response curve where gain drops off −3 dB from the mid-band value (Figure 8.16). This is why you will sometimes see selectivity referred to in terms such as "6 kHz between −3 dB points."

The IF bandwidth must be matched to the bandwidth of the received signal for best performance. If a too-wide bandwidth is selected, then the received signal will be noisy, and SNR deteriorates. If the bandwidth is too narrow, then you might experience difficulties recovering all of the information that was transmitted. For example, an AM broadcast band radio signal has audio components to 5 kHz, so

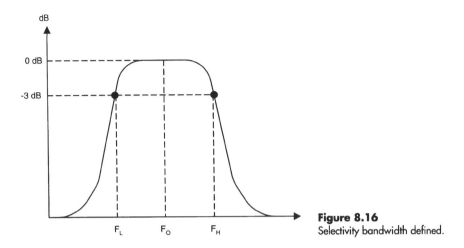

Figure 8.16
Selectivity bandwidth defined.

the signal occupies up to 10 kHz of spectrum space (F ± 5 kHz). If a 2.8 kHz SSB IF filter is selected, then it will tend to sound "mushy" and distorted.

IF Passband Shape Factor

The shape factor is a measure of the steepness of the receiver's IF passband and is taken by measuring the ratio of the bandwidth at −6 dB to the bandwidth at −60 dB. The general rule is that the closer these numbers are to each other, the better the receiver. Anything in the 1:1.5 to 1:1.9 region can be considered high quality, while anything worse than 1:3 is not worth looking at for "serious" receiver uses. If the numbers are between 1:1.9 and 1:3, then the receiver could be regarded as being middling, but useful.

The importance of shape factor is that it modifies the notion of bandwidth. The cited bandwidth (e.g., 2.8 kHz for SSB) does not take into account the effects of strong signals that are just beyond those limits. Such signals can easily "punch through" the IF selectivity if the IF passband "skirts" are not steep. After all, the steeper they are, the closer a strong signal can be without messing up the receiver's operation. Thus, selecting a receiver with a shape factor as close to the 1:1 ideal as possible will result in a more usable radio.

Distant Frequency ("Ultimate") Rejection

This specification tells something about the receiver's ability to reject very strong signals that are located well outside the receiver's IF passband. This number is stated in negative decibels (−dB), and the higher the number the better. An excellent receiver will have values in the −60 to −90 dB range, a middling receiver will have numbers in the −45 to −60 dB range, and a terrible receiver will be −44 or worse.

Stability

The *stability* specification measures how much the receiver frequency drifts as time elapses or temperature changes. The LO drift sets the overall stability of the receiver. This specification is usually given in terms of *short-term drift* and *long-term drift* (e.g., from crystal aging). The short-term drift is important in daily operation, while the long-term drift ultimately affects general dial calibration.

If the receiver is VFO controlled, or uses partial frequency synthesis (which combines VFO with crystal oscillators), then the stability is dominated by the VFO stability. In fully synthesized receivers, the stability is governed by the master reference crystal oscillator. If either an *oven-controlled crystal oscillator* (OCXO) or a *temperature-compensated crystal oscillator* (TCXO) is used for the master reference, then stability on the order of 1 part in $10^8/°C$ are achievable. For most users, the short-term stability is what is most important, especially when tuning SSB, ECSS, or RTTY signals. A common spec for a good receiver will be 50 Hz/hour after a 3-hour warm-up, or 100 Hz/hour after a 15-minute warm-up. The smaller the drift the better the receiver.

The foundation of good stability is established at design time. The local oscillator, or VFO portion of a synthesizer, must be operated in a cool, temperature-stable location within the equipment, and it must have the correct type of components. Capacitor temperature coefficients are often selected in order to cancel out temperature related drift in inductance values.

Post-design time changes can also help, but these are less likely to be possible today than in the past. The chief cause of drift problems is heat. In the days of valve oscillators, the heat of the valves produced lots of heat that created drift.

A related phenomenon seen on low-cost receivers, or certain home-brew receivers of doubtful merit, is mechanical frequency shifts. Although not seen on most modern receivers (even some very cheap designs), it was once a serious problem on less costly models. This problem is usually seen on VFO-controlled receivers in which vibration to the receiver cabinet imparts movement to either the inductor (L) or capacitor (C) element in an L-C VFO. Mechanically stabilizing these components will work wonders.

AGC Range and Threshold

Modern communications receivers must be able to handle signals over the range of about 1,000,000:1. Tuning across a band occupied by signals of wildly varying strengths is hard on the ears and hard on the receiver's performance. As a result, most modern receivers have an *automatic gain control* (AGC) circuit that smoothes out these changes. The AGC will reduce gain for strong signals and increase it for weak signals (AGC can be turned off on most HF communications receivers). The AGC range is the change of input signal (in dBμV) from some reference level (e.g., 1 μV_{EMF}) to the input level that produces a 2 dB change in output level. Ranges of 90 to 110 dB are commonly seen.

The AGC threshold is the signal level at which the AGC begins to operate. If set too low, then the receiver gain will respond to noise and irritate the user. If set too high, then the user will experience irritating shifts of output level as the band is tuned. AGC thresholds of 0.7 to 2.5 μV are common on decent receivers, with the better receivers being in the 0.7 to 1 μV range.

Another AGC specification sometimes seen deals with the speed of the AGC. Although sometimes specified in milliseconds, it is also frequently specified in subjective terms like "fast" and "slow." This specification refers to how fast the AGC responds to changes in signal strength. If set too fast, then rapidly keyed signals (e.g., CW) or noise transients will cause unnervingly large shifts in receiver gain. If set too slow, then the receiver might as well not have an AGC. Many receivers provide two or more selections in order to accommodate different types of signals.

Dynamic Performance

The dynamic performance specifications of a radio receiver are those that deal with how the receiver performs in the presence of very strong signals, either co-channel or adjacent channel. Until about the 1960s, dynamic performance was somewhat less important than static performance for most users. However, today the role of dynamic performance is probably far more critical than simplistic static performance because of crowded band conditions.

There are at least two reasons for this change in outlook. First, in the 1960s receiver designs evolved from valves to solid-state. The new solid-state amplifiers were somewhat easier to drive into nonlinearity than tube designs. Second, there has been a tremendous increase in radio frequency signals on the air. There are far more transmitting stations than ever before, and there are far more sources of electromagnetic interference (EMI—pollution of the airwaves) than in prior decades. With the advent of new and expanded wireless services available to an ever-widening market, the situation can only worsen. For this reason, it is now necessary to pay more attention to the dynamic performance of receivers than in the past.

Intermodulation Products

Understanding the dynamic performance of the receiver requires knowledge of *intermodulation products* (IP) and how they affect receiver operation. Whenever two signals are mixed together in a nonlinear circuit, a number of products are created according to $mF1 \pm nF2$, where m and n are either integers or zero. Mixing can occur in either the mixer stage of a receiver front-end, or in the RF amplifier (or any outboard preamplifiers used ahead of the receiver), if the RF amplifier is overdriven by a strong signal.

It is also theoretically possible for corrosion on antenna connections, or even rusted antenna screw terminals, to create IPs under certain circumstances. One even hears of alleged cases where a rusty downspout on a house rain gutter caused reradiated mixed signals. However, all such cases that I've heard of have that distant third- or fourth-party quality ("I know a guy whose best friend's brother-in-law saw . . . ") that suggests the profound apocryphal quality of these reports. I know of no first-hand accounts verified by a technically competent person.

The spurious IP signals are shown graphically in Figure 8.17. Given input signal frequencies of F1 and F2, the main IPs are:

Second-order:	$F1 \pm F2$
Third-order:	$2F1 \pm F2$
	$2F2 \pm F1$
Fifth-order:	$3F1 \pm 2F2$
	$3F2 \pm 2F1$

The second-order and third-order products are those normally specified in a receiver because they tend to be the strongest. In general, even-order IMD prod-

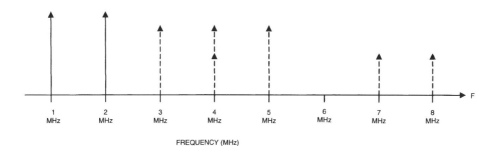

FREQUENCY (MHz)

A

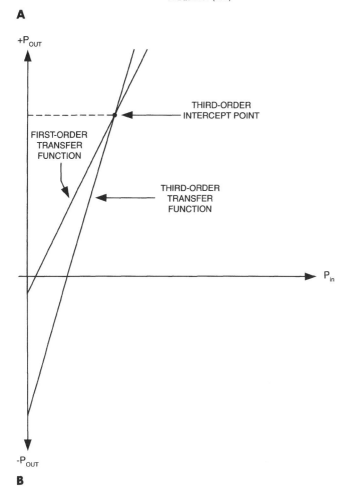

B

Figure 8.17
(A) Intermodulation products;
(B) Third-order and first-order products compared.

ucts (2, 4, etc.) tend to be less of a problem because they can often be ameliorated by using external filtering ahead of the receiver's antenna input. Prefiltering ("pre-selection") tends to reduce the amplitude of out-of-channel interfering signals, reducing the second-order products within the channel. Third-order IMD products are more important because they tend to reflect on the receiver's dynamic range, as well as its ability to handle strong signals. The third-order products are usually not easily influenced by external filtering.

When an amplifier or receiver is overdriven, the second-order content of the output signal increases as the square of the input signal level, while the third-order responses increase as the cube of the input signal level. When expressed in dB the third-order transfer function has a slope three times that of the first-order transfer function (Figure 8.17B).

Consider the case where two HF signals, F1 = 10 MHz and F2 = 15 MHz, are mixed together. The second-order IPs are 5 and 25 MHz; the third-order IPs are 5, 20, 35, and 40 MHz; and the fifth-order IPs are 0, 25, 60, and 65 MHz. If any of these are inside the passband of the receiver, then they can cause problems. One such problem is the emersion of "phantom" signals at the IP frequencies. This effect is seen often when two strong signals (F1 and F2) exist and can affect the front-end of the receiver, and one of the IPs falls close to a desired signal frequency, F_d. If the receiver were tuned to 5 MHz, for example, a spurious signal would be found from the F1-F2 pair given above.

Another example is seen from strong in-band, adjacent channel signals. Consider a case where the receiver is tuned to a station at 9610 kHz, and there are also very strong signals at 9600 kHz and 9605 kHz. The near (in-band) IP products are:

Third-order:	9595 kHz (ΔF = 15 kHz)
	9610 kHz (ΔF = 0 kHz)[ON CHANNEL!]
Fifth-order:	9590 kHz (ΔF = 20 kHz)
	9615 kHz (ΔF = 5 kHz)

Note that one third-order product is on the same frequency as the desired signal and could easily cause interference if the amplitude is sufficiently high. Other third- and fifth-order products may be within the range where interference could occur, especially on receivers with wide bandwidths.

The IP orders are theoretically infinite because there are no bounds on either m or n. However, in practical terms, because each successively higher order IP is reduced in amplitude compared with its next lower-order mate, only the second-order, third-order, and fifth-order products usually assume any importance. Indeed, only the third-order is normally used in receiver specification sheets.

−1 dB Compression Point

An amplifier produces an output signal that has a higher amplitude than the input signal. The transfer function of the amplifier (indeed, any circuit with output and input) is the ratio OUT/IN, so for the power amplification of a receiver RF amplifier it is P_o/P_{in} (or, in terms of voltage, V_o/V_{in}). Any real amplifier will saturate given a strong enough input signal (see Figure 8.18). The sloping dotted line in Figure 8.18 represents the theoretical output level for all values of input signal (the slope of the line represents the gain of the amplifier). As the amplifier saturates (solid line), however, the actual gain begins to depart from the theoretical at some level of input signal (P_{in1}). The −1 dB compression point is that output level at which the actual gain departs from the theoretical gain by −1 dB.

The −1 dB compression point is important when considering either the RF amplifier ahead of the mixer (if any), or any outboard preamplifiers that are used.

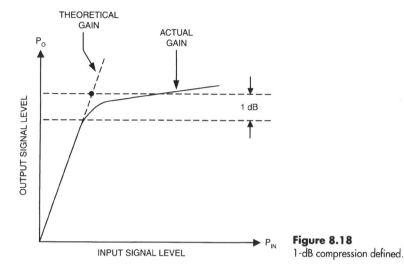

Figure 8.18
1-dB compression defined.

The −1 dB compression point is the point at which intermodulation products begin to emerge as a serious problem. It is also the case that harmonics are generated when an amplifier goes into compression. A sine wave is a "pure" signal because it has no harmonics (all other waveshapes have a fundamental frequency plus harmonic frequencies). When a sine wave is distorted, however, harmonics arise. The effect of the compression phenomenon is to distort the signal by clipping the peaks, thus raising the harmonics and intermodulation distortion products.

Third-Order Intercept Point

It can be claimed that the third-order intercept point (TOIP) is the single most important specification of a receiver's dynamic performance because it predicts the performance as regards intermodulation, crossmodulation, and blocking desensitization.

Third-order (and higher) intermodulation products (IP) are normally very weak and do not exceed the receiver noise floor when the receiver is operating in the linear region. But as input-signal levels increase, forcing the front-end of the receiver toward the saturated nonlinear region, the IP emerges from the noise (Figure 8.19) and begins to cause problems. When this happens, new spurious signals appear on the band and self-generated interference begins to arise.

Figure 8.20 shows a plot of the output signal versus fundamental input signal. Note the output compression effect that was seen earlier in Figure 8.18. The sloping dotted gain line continuing above the saturation region shows the theoretical output that would be produced if the gain did not clip. It is the nature of third-order products in the output signal to emerge from the noise at a certain input level, and to increase as the cube of the input level. Thus, the slope of the third-order line increases 3 dB for every 1-dB increase in response to the fundamental signal. Although the output response of the third-order line saturates similarly to that of the fundamental signal, the gain line can be continued to a point where it

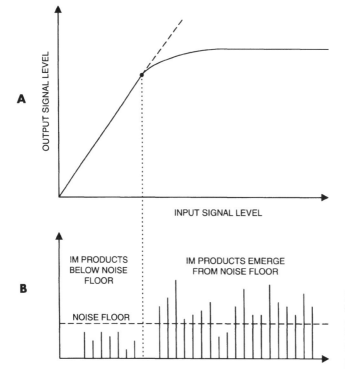

Figure 8.19
Operating above compression point causes IM products to emerge from the noise floor. (A) Output-vs.-input signal levels; (B) Distortion products and noise floor.

intersects the gain line of the fundamental signal. This point is the *third-order intercept point* (TOIP).

Interestingly enough, one receiver feature that can help reduce IP levels back down under the noise is the use of a *front-end attenuator*, also known as an *input attenuator*. Even a few dB of input attenuation is often enough to cause the IPs to drop back into the noise, while afflicting the desired signals only a small amount.

Other effects that reduce the overload caused by a strong signal also help. The apparent third-order performance of a receiver can improve dramatically when either a lower-gain antenna or front-end attenuator is used. Inserting a 6-dB barrel attenuator in the input ("antenna") line can eliminate or reduce the IP products, showing just the actual signals. Rotating a directional antenna away from the direction of the interfering signal will also accomplish this effect in many cases.

Preamplifiers are popular receiver accessories, but they can often reduce rather than enhance performance. Two problems commonly occur (assuming the preamp is a low-noise device; if not, there are three). The best known problem is that the preamp amplifies noise as much as signals, and while it makes the signal louder it also makes the noise louder by the same amount. Since it is the signal-to-noise ratio that is important, one does not improve the situation. Indeed, if the preamplifier is itself noisy, it will deteriorate the SNR. The other problem is less well known, but potentially more devastating. If the increased signal levels applied to the receiver drive the receiver in a nonlinearly way, then IPs begin to emerge. One source reported an event where transmissions were being heard at several spots on

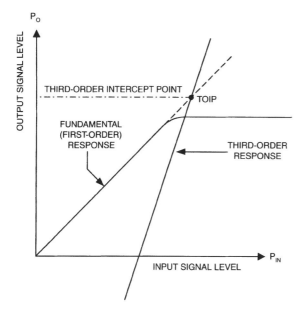

Figure 8.20
Third-order intercept point (TOIP or IP3) defined.

the band. It was discovered that the complainant was using two preamplifiers in cascade to achieve higher gain, and that disconnecting them caused the spurs to evaporate. This was clearly a case of a preamplifier deteriorating, rather than improving, a receiver's performance.

When evaluating receivers, a TOIP of +5 to +20 dBm is excellent performance, while up to +27 dBm is relatively easily achievable, and +35 dBm has been achieved with good design; anything greater than +50 dBm is close to miraculous (but attainable). Receivers are still regarded as good performers in the 0 to +5 dBm range, and middling performers in the −10 to 0 dBm range. Anything below −10 dBm is not a very wonderful a machine to own. A general rule is to buy the best third-order intercept performance that you can afford, especially if there are strong signal sources in your vicinity.

Dynamic Range

The *dynamic range* of a radio receiver is the range from the minimum discernible signal to the maximum allowable signal (measured in decibels, dB). While this simplistic definition is easy to understand conceptually, in the concrete it is a little more complex. Several definitions of dynamic range are used.

One definition of dynamic range is that it is the input signal difference between the sensitivity figure (e.g., 0.5 μV for 10 dB S + N/N) and the level that drives the receiver far enough into saturation to create a certain amount of distortion in the output. This definition was common on consumer broadcast band receivers at one time (especially automobile radios, where dynamic range was somewhat more important due to mobility). A related definition takes the range as the distance in dB from the sensitivity level and the −1 dB compression point. Still another definition, the *blocking dynamic range,* is the range of signals from the sensitivity level to the blocking level (see below).

A problem with the above definitions is that they represent single-signal cases, so they do not address the receiver's dynamic characteristics. Dye (1993) provides both a "loose" and a more formal definition that is somewhat more useful and is at least standardized. The loose version is that dynamic range is the range of signals over which dynamic effects (e.g., intermodulation) do not exceed the noise floor of the receiver. Dye's recommendation for HF receivers is that the dynamic range is two-thirds the difference between the noise floor and the third-order intercept point in a 3-kHz bandwidth. Dye also states an alternative: dynamic range is the difference between the fundamental response input signal level and the third-order intercept point along the noise floor, measured with a 3-kHz bandwidth. For practical reasons, this measurement is sometimes made not at the actual noise floor (which is sometimes hard to ascertain), but rather at 3 dB above the noise floor.

There is a measurement procedure that produces similar results (the same method is used for many amateur radio magazine product reviews). Two equal-strength signals are input to the receiver at the same time. The frequency difference has traditionally been 20 kHz for HF and 30 to 50 kHz for VHF receivers (modern band crowding may indicate a need for a specification at 5 kHz separation on HF). The amplitudes of these signals are raised until the third-order distortion products are raised to the noise floor level.

For 20-kHz spacing, using the two-signal approach, anything over 90 dB is an excellent receiver, while anything over 80 dB is at least decent.

The difference between the single-signal and two-signal (dynamic) performance is not merely an academic exercise. Besides the fact that the same receiver can show as much as a 40-dB difference between the two measures (favoring the single-signal measurement), the most severe effects of poor dynamic range show up most in the dynamic performance.

Blocking

The blocking specification refers to the ability of the receiver to withstand very strong off-tune signals that are at least 20 kHz away from the desired signal, although some use 100-kHz separation. Very strong signals appearing at the input terminals of a receiver may desensitize the receiver, that is, reduce the apparent strength of desired signals over what they would be if the interfering signal were not present.

Figure 8.21 shows the blocking behavior. When a strong signal is present, it takes up more of the receiver's resources than normal, so there is not enough of the output power budget to accommodate the weaker desired signals. But if the strong undesired signal is turned off, then the weaker signals receive a full measure of the unit's power budget.

The usual way to measure blocking behavior is to input two signals: a desired signal at 60 dBμV, and another signal 20 (or 100) kHz away at a much stronger level. The strong signal is increased to the point where blocking desensitization causes a 3-dB drop in the output level of the desired signal. A good receiver will show ≥90 dBμV, with many being considerably better. An interesting note about

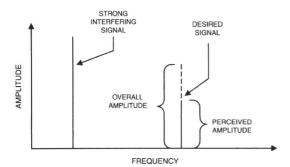

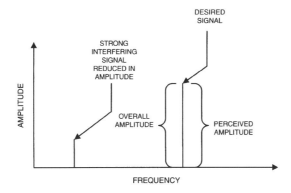

Figure 8.21
Desensitization due to strong interfering signal can be overcome if the signal levels are reduced below compression point.

modern receivers is that their blocking performance is so good that it is often necessary to specify the input level difference (dB) that causes a 1-dB drop, rather than a 3-dB drop, of the desired signal's amplitude.

The phenomenon of blocking leads us to an effect that is often seen as paradoxical on first blush. Many receivers are equipped with front-end attenuators that permit fixed attenuation values of 1 dB, 3 dB, 6 dB, 12 dB, or 20 dB (or some subset) to be inserted into the signal path ahead of the active stages. When a strong signal that is capable of causing desensitization is present, *adding attenuation often increases the level of the desired signals in the output,* even though overall gain is reduced. This occurs because the overall signal that the receiver front-end is asked to handle is below the threshold where desensitization occurs.

Crossmodulation

Crossmodulation is an effect in which amplitude modulation (AM) from a strong undesired signal is transferred to a weaker desired signal. Testing is usually done (in HF receivers) with a 20-kHz spacing between the desired and undesired signals, a 3-kHz IF bandwidth on the receiver, and the desired signal set to 1,000 μV_{EMF} (−53 dBm). The undesired signal (20 kHz away) is amplitude modulated to the 30% level. This undesired AM signal is increased in strength until an unwanted AM output 20 dB below the desired signal is produced. A crossmodulation specification ≥100 dB would be considered decent performance. This figure is often not given for modern HF receivers, but if the receiver has a good third-order intercept point, then it is likely also to have good crossmodulation performance.

Crossmodulation is also said to occur naturally, especially in transpolar and North Atlantic radio paths where the effects of the aurora are strong. According to one (possibly apocryphal) legend, there was something called the "Radio Luxembourg Effect" discovered in the 1930s. Modulation from very strong broadcasters appeared on the Radio Luxembourg signal received in North America. This effect was said to be an Ionospheric crossmodulation phenomenon. If you or anyone you know has any direct experience with this effect, or a literature citation, I would be interested in hearing from you.

Reciprocal Mixing

Reciprocal mixing occurs when noise sidebands from the local oscillator (LO) signal in a superheterodyne receiver mix with a strong undesired signal that is close to the desired signal. Every oscillator signal produces noise, and that noise tends to amplitude modulate the oscillator's output signal. It will thus form sidebands on either side of the LO signal. The production of phase noise in all LOs is well known, but in more recent designs the digitally produced synthesized LOs are prone to additional noise elements as well. The noise is usually measured in -dBc (decibels below carrier, or, in this case, dB below the LO output level).

In a superheterodyne receiver, the LO beats with the desired signal to produce an intermediate frequency (IF) equal to either the sum (LO + RF) or difference (LO − RF). If a strong unwanted signal is present, then it might mix with the noise sidebands of the LO, to reproduce the noise spectrum at the IF frequency (see Figure 8.22). In the usual test scenario, the reciprocal mixing is defined as the level of the unwanted signal (dB) at 20 kHz required to produce a noise sidebands 20 down from the desired IF signal in a specified bandwidth (usually 3 kHz on HF receivers). Figures of −90 dBc or better are considered good.

The importance of the reciprocal mixing specification is that it can seriously deteriorate the observed selectivity of the receiver, yet it is not detected in the nor-

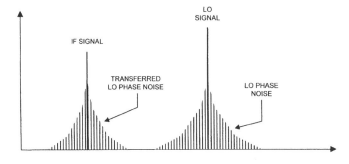

Figure 8.22
LO phase noise can be transferred to IF signal by reciprocal mixing.

mal static measurements made of selectivity (it is a "dynamic selectivity" problem). When the LO noise sidebands appear in the IF, the distant frequency attenuation (>20 kHz off-center of a 3-kHz bandwidth filter) can deteriorate 20 to 40 dB.

The reciprocal mixing performance of receivers can be improved by eliminating the noise from the oscillator signal. Although this sounds simple, in practice it is often quite difficult. A tactic that will work well, at least for those designing their own receiver, is to add high-Q filtering between the LO output and the mixer input. The narrow bandwidth of the high-Q filter prevents excessive noise sidebands from getting to the mixer. Although this sounds like quite an easy solution, as they say, "the devil's in the details."

IF Notch Rejection

If two signals fall within the passband of a receiver they will both compete to be heard. They will also heterodyne together in the detector stage, producing an audio tone equal to their carrier frequency difference. For example, suppose we have an AM receiver with a 5-kHz bandwidth and a 455-kHz IF. If two signals appear on the band such that one appears at an IF of 456 kHz and the other is at 454 kHz, then both are within the receiver passband and both will be heard in the output. However, the 2-kHz difference in their carrier frequency will produce a 2-kHz heterodyne audio-tone difference signal in the output of the AM detector.

In some receivers, a tunable, high-Q (narrow and deep) notch filter is in the IF amplifier circuit. This tunable filter can be turned on and then adjusted to attenuate the unwanted interfering signal, reducing the irritating heterodyne. Attenuation figures for good receivers vary from −35 to −65 dB, or so (the more negative the better).

There are some trade-offs in notch filter design. First, the notch filter Q is more easily achieved at low IF frequencies (such as 50 to 500 kHz) than at high IF frequencies (e.g., 9 MHz and up). Also, the higher the Q the better the attenuation of the undesired squeal, but the touchier it is to tune. Some happy middle ground between the irritating squeal and the touchy tune is mandated here.

Some receivers use audio filters rather than IF filters to help reduce the heterodyne squeal. In the AM broadcast band, channel spacing is typically 10 kHz and the transmitted audio bandwidth (hence the sidebands) are 5 kHz. Designers of AM BCB receivers usually insert an R-C low-pass filter with a −3 dB point just above 4 or 5 kHz right after the detector in order to suppress the 10 kHz heterodyne. This R-C filter is called a "tweet filter" in the slang of the electronic service/repair trade.

Another audio approach is to sharply limit the bandpass of the audio amplifiers. For AM BCB reception, a 5-kHz bandpass is sufficient, so the higher frequencies can be rolled off at a fast rate in order to produce only a small response an octave higher (10 kHz). In shortwave receivers, this option is weaker because the station channels are typically 5 kHz, and many don't bother to honor the official channels anyway. On the amateur radio bands, frequency selection is a perpetually changing "ad-hocracy" at best. Although the shortwave bands typically only need 3-kHz bandwidths for communications, and 5 kHz for broadcast, the tweet filter and audio roll-off might not be sufficient. In receivers that lack an effective IF notch filter, an audio notch filter can be provided. This accessory can even be added after the fact (as an outboard accessory) once you own the receiver.

Internal Spurii

All receivers produce a number of internal spurious signals that sometimes interfere with the operation. Both old and modern receivers have spurious signals from assorted high-order mixer products, from power supply harmonics, parasitic oscillations, and a host of other sources. Newer receivers with either (or both) synthesized local oscillators and digital frequency readouts produce noise and spurious signals in abundance. (Note: low-power digital chips with slower rise times—CMOS, NMOS, etc.—are generally much cleaner than higher-power, fast-rise time chips like TTL devices.)

With appropriate filtering and shielding, it is possible to hold the "spurs" down to −100 dB relative to the main maximum signal output, or within about 3 dB or the noise floor, whichever is lower.

Harry Helms, a writer of shortwave books (1994) has several high-quality receivers, including valve models and modern synthesized models. His comparisons of basic spur/noise levels were something of a surprise. A high-quality valve receiver from the 1960s appeared to have a lower noise floor than the modern receivers. Helms attributed the difference to the internal spurii from the digital circuitry used in the modern receivers.

RECEIVER SENSITIVITY MEASUREMENTS

Radio receiver specifications can be verified using test equipment and a few simple procedures. Such tests are made to evaluate receivers, to troubleshoot problems, and to verify performance. A number of receiver parameters are important, but perhaps the one that is most commonly discussed is *sensitivity*. Let's take a look at how these tests are done.

Sensitivity

Receiver sensitivity is a measure of how well the receiver will pick up very weak signals. As with most engineering measurements, the notion of sensitivity is an operational definition. In other words, there are standard procedures that will yield coherent results by which different receivers (or the same receiver before and after repairs) can be compared.

Sensitivity is basically a game of *signal-to-noise ratio* (SNR), or more properly, the signal-plus-noise-to-noise ratio (S + N/N). For every receiver or amplifier there is a basic noise level consisting of the noise produced external to the receiver and noise produced inside the receiver. Even a receiver with its antenna input terminated in a shielded matching resistor, rather than an antenna or signal generator, will show a certain amount of thermal noise.

One important consideration when making sensitivity measurements (or comparing receiver sensitivity specifications) is bandwidth. Thermal and other forms of noise are gaussian and distributed over all possible bandwidths. The value of the noise at any given instant is dependent on the bandwidth of the channel. For most receivers this means the IF selectivity bandwidth, although in some cases the audio bandwidth is less than the IF so that number would dominate.

Figure 8.23 depicts two different definitions of SNR. Basically, you can't hear signals down in the noise. The minimum discernible signal (MDS) is operationally defined as the signal level that is the same as the noise floor, or the signal level that is 3 dB above the receiver's noise floor. But that sensitivity is not all that useful for most applications. There may be people out there who can listen to a signal that is only 3 dB above the noise floor. Most people, however, require a higher SNR to be practical. Although some definitions of SNR use 6, 12, or 20 dB, the standard for practical sensitivity is that it is the signal level that produces a 10-dB SNR. This definition is found on most CW, AM, and SSB receivers.

Signal Generator

The signal generator selected to make sensitivity measurements must have very high isolation figures. Most "service grade" signal generators are useful for doing troubleshooting, but are not satisfactory for making sensitivity measurements. The reason is that signal escapes around the cabinet flanges and control bushings. If you have a sensitive receiver or spectrum analyzer, then you can detect this signal.

Want to give it a try? Connect a shielded dummy load to the output of the signal generator, and turn the signal generator's output down to zero. Connect a whip or wire antenna to the receiver's antenna input, and then tune the receiver across the signal generator frequency with the RF gain cranked all the way up.

The signal generator should also have a calibrated output control. The correct calibrations are either *dBm* (power decibels relative to one milliwatt in a 50-ohm load), or microvolts (μV). Some signal generators have an output meter that can set relative output, but can become "calibrated" if a calibrated step-attenuator is connected between the output of the signal generator and the receiver under test.

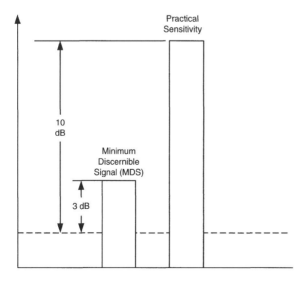

Figure 8.23
Practical sensitivity compared with minimum discernible signal (MDS).

You can find the exact level if you can measure the high level output of the signal generator.

"Laboratory grade" signal generators may be beyond the means of many people, but there is a relatively vigorous market in used or surplus equipment. There are a number of sources of such equipment listed on the World Wide Web. If you don't need the latest digitally synthesized signal generators, then you will be able to find good signal generators at low cost.

Test Set-up

Figure 8.24 shows the test set-up for most receiver sensitivity measurements. The attenuator is optional and may not be needed if the signal generator is adequately equipped with a good-quality calibrated output attenuator. When measuring an AM receiver, set the signal generator modulation for 30% depth and 1,000 Hz.

The receiver output level is measured using an audio AC voltmeter. Ideally, the instrument should be calibrated in decibels as well as volts and should have RMS reading capability.

The receiver must be correctly set up or the measurement will be in error. In most test set-ups, the receiver's RF and AF gain controls are turned to maximum, and the squelch is turned off. Furthermore, the *automatic gain control* (AGC) must be either turned off, or in the case of some models, clamped with a DC level according to the manufacturer's directions.

Minimum Discernible Signal (MDS) Sensitivity

To make the MDS measurement we need to find the signal level in dBm or μV that is 3 dB above the receiver noise floor.

1. Connect the equipment as in Figure 8.24, and set the receiver and signal generator to the same frequency.
2. Turn the signal generator output all the way down to zero.
3. Crank the RF gain and AF controls all the way up (you may want to set the audio output control to a convenient level if you don't have a dummy speaker load).
4. To make the measurement, you first measure the RMS value of the noise ("hiss") output on the AC voltmeter, and then increase the signal-generator output level until the receiver output level increases 3 dB.

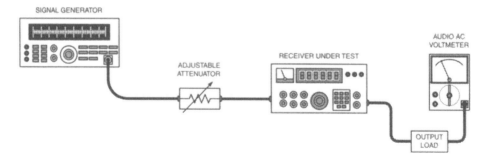

Figure 8.24 Standard test set-up for sensitivity measurements.

You can also determine the numerical value of the receiver noise floor by the same approach. Measure the output noise level, and then find the MDS by the procedure above. The receiver noise floor level will be the same as the signal generator output level (less any attenuation in line).

Standard Output Conditions

A sensitivity specification used for consumer radio receivers uses a standard output approach. A typical receiver sensitivity spec might read "xx μV for 400-mW in an 8-ohm load when modulated 30% by 1-kHz." The same equipment set up (Figure 8.24) can be used for this measurement. A power of 400 mW (0.4-watt) in an 8-ohms load is the same as 1.789 volts RMS, which can be read on the AC voltmeter. Use an 8-ohms noninductive resistor for the load rather than the loudspeaker-otherwise the sound levels are pretty annoying. Adjust the signal generator output level for an RMS output voltage of 1.789 volts, and read the output level from the signal-generator controls.

Full-Power Sensitivity

Some older receivers use the full-power sensitivity figure. This is the signal level that will produce the full-rated audio output power. Set the signal generator for 1-kHz modulation with 30% depth. Tune the radio and signal generator to the same frequency, and crank up the output until the audio output power is at the full-rated power level (e.g., 400 mW, 1 watt, etc.). The signal level that produces this condition is the "sensitivity" of the receiver.

10-dB (S + N)/N Test

The 10-dB test method is the same as the 3-dB MDS method, except that the signal generator level is increased until the output is 10 dB above the noise floor.

An alternate method is sometimes used on AM receivers:

1. Set up the signal generator and receiver as discussed above.
2. Set the output of the receiver to produce at least 0.5 watts audio output, or if the rated output power is lower than 1 watt set it for at least 50-mW audio output power.
3. Turn off the modulation. If the audio output drops at least 10-dB (or more), then the signal generator setting is at the 10-dB S + N/N level. If the level drops less than 10-dB, then readjust the signal-generator output level upward a small amount and try again.

On-site Effective Sensitivity Test

This test is only done on-site where the receiver is installed. It is intended to get some idea of how well the receiver performs in its actual installed environment. Figure 8.25 shows the test set-up.

1. Measure the 10-dB S + N/N sensitivity as discussed above (see Figure 8.24 for set-up), and write down the number.
2. Connect the hybrid combiner, two-position coaxial switch, antenna, and dummy load into the circuit.
3. Set the switch to the dummy load and measure the sensitivity. It will be considerably worse than the 10-dB sensitivity.
4. Set the coaxial switch to the antenna, and again measure the 10-dB sensitivity. It should be lower still.

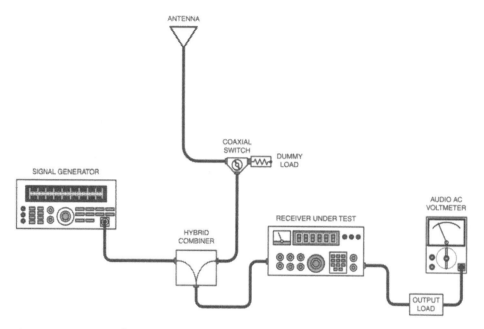

Figure 8.25 On-site effective sensitivity test.

The effective sensitivity is $SNR_{10dB} - (SNR_{Load} - SNR_{ANT})$. The figure $SNR_{Load} - SNR_{ANT}$ is the degradation factor. For example, suppose the 10-dB SNR is -122 dBm, the SNR when the load is connected is -77 dBm, and when the antenna is connected it is -70 dBm. The effective on-site SNR is:

$$SNR_{EFF} = SNR_{10dB} - (SNR_{Load} - SNR_{ANT})$$
$$SNR_{EFF} = -122 \text{ dBm} - [(-77 \text{ dBm}) - (-70 \text{ dBm})]$$
$$SNR_{EFF} = -122 \text{ dBm} - [-7 \text{ dBm}] = -115 \text{ dBm}$$

The effective sensitivity is only valid for the given site and conditions present when the test is performed. If the site is changed, or if the noise generators and other signals present change, then the test must be repeated.

FM Receiver Sensitivity

There are two basic methods for measuring the sensitivity of FM receivers: *20-dB quieting* and *12-dB SINAD.* The 20-dB quieting method is typically used on FM broadcast band receivers, and was once popular for communications receivers as well. More recently, the 12-dB SINAD method is preferred.

20-dB Quieting Method

This method relies on the fact that the FM detector will suppress noise once the limiting signal level is reached. The well-known capability of FM to eliminate noise relies on the fact that most noise amplitude modulates the carrier. If the amplitude can be clamped below the level where the noise is effective, then the frequency variations can be detected to recover the audio. This effect is called "quieting," which is the reduction of noise as the signal level increases.

To measure the 20-dB quieting sensitivity:

1. Connect the receiver and signal generator as in Figure 8.24. Keep the signal-generator output at zero. The modulation (deviation) should be set to whatever is appropriate for the class of receiver being measured.
2. Turn the RF gain all the way up. Set the audio output to produce a convenient reading in the high end of the AC voltmeter scale.
3. Measure the output noise level and write it down.
4. Turn the signal generator output level up until the reading on the AC voltmeter drops 20 dB. The signal generator output level that accomplishes this is the 20-dB quieting sensitivity (typically less than 1 μV).

SINAD Sensitivity

The sensitivity of FM receivers is often expressed in terms of SINAD. This approach to signal-to-noise ratio recognizes that the problem of detection depends on not simply signal and noise level, but also distortion. The SINAD (signal-noise-distortion) method is described by Equation 8.12.

$$SINAD = \frac{Signal + Noise + Distortion}{Noise + Distortion} \qquad [8.12]$$

In terms of decibels, the following equation is used:

$$SINAD\ (dB) = 20\ LOG \left[\frac{V_{Signal} + V_{Noise} + V_{Distortion}}{V_{Noise} + V_{Distortion}} \right] \qquad [8.13]$$

Where:

SINAD(dB) is the SINAD sensitivity expressed in decibels (dB)

V_{Signal} is the output voltage due to signal

V_{Noise} is the output voltage due to noise

$V_{Distortion}$ is the output voltage due to distortion

The standard 12-dB SINAD sensitivity corresponds to a 4:1 SNR ratio, in which the sum or noise and distortion is 25% of the signal voltage. As signal levels get higher, the SINAD and 10-dB S/N values tend to converge.

Figure 8.26 shows a typical test set-up for the SINAD measurement. The output AC voltmeter is augmented by a Total Harmonic Distortion (THD) analyzer,

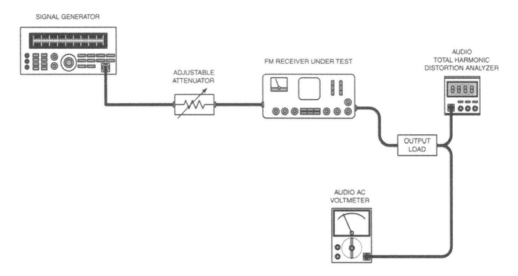

Figure 8.26   SINAD sensitivity test set-up.

both of which measure the output signal across the audio load (speaker, load resistor, etc.)

1. Set the signal generator frequency and receiver frequency to the same value.
2. Set standard conditions: modulating frequency 1-kHz sine wave; deviation set to 60% of the peak deviation used for that service. For example, for an FM BCB receiver, deviation is ±75 kHz, so set the signal generator deviation to $0.6 \times \pm 75$ kHz = ±45 kHz. For a communications receiver designed for ±5 kHz deviation, set deviation to $0.6 \times \pm 5$ kHz = ±3 kHz.
3. Adjust the receiver audio output to approximately 50% of the receiver's rated audio output.
4. Adjust the signal generator output until the input signal is high enough to produce 25% distortion. This is the 12-dB SINAD sensitivity.

Special SINAD sensitivity meters are available that combine the THD analyzer and audio voltmeter functions in one instrument.

RECEIVER NOISE MEASUREMENT

Radio reception is a game of signal-to-noise ratio. For that reason, it is sometimes necessary to measure noise levels and the noise performance of receivers.

Noise Floor

The noise floor of a receiver is the same as the minimum detectable signal (MDS) and is measured in exactly the same way. When you measure the MDS, it is defined as the signal level that causes an output 3 dB above the noise floor. Hence, when you measure the MDS, you have also measured the noise floor.

Setting standard conditions is necessary to make this measurement properly. For example, one caution is to use a signal level that is at least 10 dB above the expected sensitivity of the receiver, and then work down to find the 3 dB increase level.

One of the standard conditions is to ensure that the noise floor measurements are made in a standard bandwidth. It is often the case that receivers with multiple bandwidths will specify noise floor and sensitivity in the narrowest available bandwidth. However, that may not be useful if the normal mode for that receiver requires a wider bandwidth. For example, an H.F. communications receiver may have filters for AM mode (BW = 6 kHz), single-sideband (BW = 2.8 kHz), and CW (BW = 500 Hz). If the mode required for your application is SSB, then do not rely on sensitivity and noise floor measurements made on the 500-Hz bandwidth for the CW mode.

1-dB Compression Point

The 1-dB compression point is the input signal level at which receiver gain drops 1-dB (Figure 8.27). The gain of the system is depicted by the P_{out}/P_{in} line. This characteristic is linear up to a point, that is, a 1-dB increase in input signal level causes a proportionally scaled output level change. At some point, however, the receiver is saturated and cannot accommodate any further input signal. The operational definition of where this occurs is the 1-dB compression point.

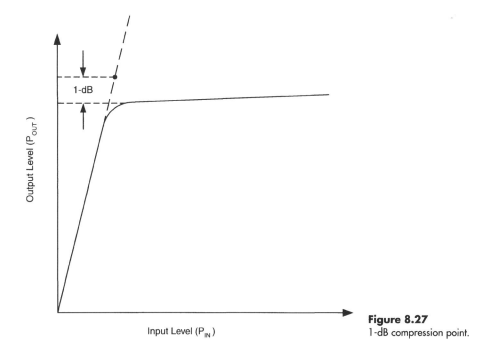

Figure 8.27
1-dB compression point.

To measure this point, use the standard test set-up of Figure 8.24. Bring the input power level applied to the receiver antenna terminals up from some low value in 1-dB steps until the receiver output level drops 1 dB. The input level at which this occurs is the 1-dB compression point.

Dynamic Range

Dynamic range represents the total range of input signal levels that can be accommodated. The classical dynamic range measure is the *blocking dynamic range* (BDR). There is also the *Third-Order IMD Dynamic Range* (TOIMDDR).

Blocking Dynamic Range

The BDR measures the difference between the receiver noise floor and the level of an off-channel undesired signal to reduce the sensitivity to on-channel signals by a specified amount (Figure 8.28). In other words, it is a measure of the range from MDS to a specified *desensitization* of the receiver.

Connect the test set-up of Figure 8.29. Two signal generators are coupled through an isolating hybrid to a step attenuator and then to the receiver. An AC audio voltmeter is used to measure the output level of the receiver. Figure 8.28 shows the signal situation for the receiver input. Frequency F1 is the desired signal, while F2 is the interfering signal.

The amplitude of F1 should be set such that it is above the receiver MDS by at least the amount of the required minimum S/N ratio (e.g., 10 dB), but below the point where it begins to increase IMD products. Using a higher level minimizes the noise error contribution. The exact level is somewhat test-procedure dependent. If the receiver automatic gain control (AGC) can be disabled, then set the signal level about 10 dB below the 1-dB compression point. If the AGC cannot be disabled, then the signal level should be lower, such as about 20 dB above the noise floor.

Frequencies F1 and F2 must have some specified spacing (ΔF). In most cases, an H.F. receiver will call for a 20-kHz spacing, while a VHF/UHF receiver will call for a 20-kHz or 100-kHz spacing. Some special-purpose microwave receivers sometimes look at considerably larger spacing. Whatever the case, the same spac-

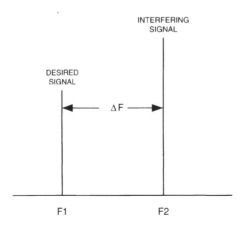

Figure 8.28
ΔF is the difference between a desired signal and an undesired signal.

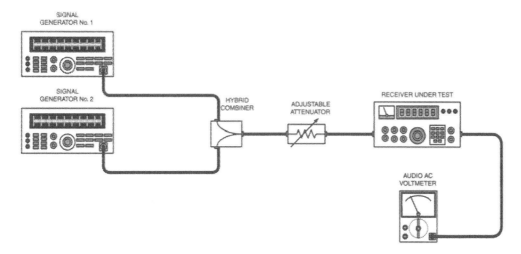

Figure 8.29 IM distortion test set-up.

ing should be used on different receivers when comparisons are made, because it seriously affects the results.

At the start of the measurement the F1 is turned on, F2 is turned off, and the receiver is adjusted for maximum output of F1. Turn on F2 at some level such as −100 dBm. Increase the level of F2 in 1-dB steps until the output level of the receiver drops 1-dB. The level at which this occurs represents the blocking dynamic range.

Perform this test on automatic test equipment, then the output levels will be automatically stepped. Also, in some cases the frequencies will also be stepped up and down the band, although maintaining the frequency separation ΔF. Adequate *settling time* between measurements must be programmed into the protocol. The signal-generator setting time rarely dominates this case, but be rather wary of the AGC settling time. By their nature, AGC circuits have a time constant, and it might be a relatively long period (e.g., 5 seconds). Understand the AGC time constant before programming a frequency or signal-level stepping protocol.

Third-Order IMD Dynamic Range (TOIMDDR)

A test set-up similar to that for BDR is used. Set the signal generators for a convenient output level such as −20 dBm, and the frequencies in-band with a specified ΔF spacing. The attenuator is set for a high degree of attenuation (low-signal level). Increase the signal by decreasing attenuation in 1-dB steps until a third-order response appears in the output. The actual signal level applied to the receiver can, as usual, be calculated from the signal generator and attenuator settings. The difference between the MDS and the level that is determined by this procedure is the TOIMDDR.

A Cautionary Note

In both this measurement and the IMD measurement (see below) it is important to ensure that only the receiver IMD products are being measured. Anytime a receiver or amplifier is overdriven into a nonlinear operating region, the possibility exists for the creation of unwanted distortion products. Even a single signal can produce first-order effects. If, for example, frequency F is applied to the input in sufficient power to overload the front-end, then harmonics at 2F, 3F, and so forth might be generated. This is also the case with normal mF1 ±nF2 IMD products higher than first-order.

Several mechanisms exist, and must be addressed. First, it is possible that the hybrid coupler will be nonlinear. This can occur when broadband ferrite or powdered iron-core transformers are used in the hybrid coupler. The coupler must produce very low levels of IMD, or it will affect the results of the test. Typically, a third-order intercept point (also known as TOIP or IP3) of > 50 dBm is required of the coupler. If the hybrid coupler being used has a lower TOIP, or if the receiver has a particularly high TOIP (> 45 dBm), then do not use any test configuration that places attenuation between the receiver input and the output of the coupler. That configuration forces the coupler to operate at too high a level.

Another scenario is a signal from one signal generator entering the output circuits of the other. Most amplifiers produce IMD products when signals arrive from the outside. This problem is reduced to negligible levels if the signal generator outputs have a high degree of inherent isolation, if attenuation is used between the coupler and the signal generators, and if the input-to-input port isolation of the hybrid is high. Try to get as much isolation as possible between signal sources, and between the sources and the receiver (> 90 dB). It is also prudent to use a signal generator with a high-level output (i.e., ≈ +15 to +22 dBm).

Another problem with coupling between signal generators is that many high-quality signal generators employ a feedback-controlled automatic level control (ALC) that samples and rectifies the RF output signal, and then feeds back the derived DC control voltage to an amplitude modulator. When external signals arrive, they can affect the ALC in two ways. First, the extra signal level may influence feedback levels. Second, and more likely, the IMD products will get into the feedback system, causing beat notes that modulate the regular output signal.

> Tidbit: How do you tell if the IMD is due to the receiver or the test fixture? In most cases, reducing the signal levels will tell the tale. If the ratio of the desired to IMD responses changes when the input signal level to the receiver is changed, then it is a reasonable assumption that the IMD is generated inside the receiver. If the ratio does not change, then it is probably due to the test fixture.

Intermodulation Distortion

Intermodulation distortion (IMD) occurs when two frequencies, F1 and F2, mix together to produce heterodyne products that were not in the original set of signals. Figure 8.30 shows this behavior. When F1 and F2 are sufficiently strong, the receiver becomes nonlinear, so mixing will occur. When this happens IM products rise up out of the noise.

There are several different methods for measuring the IMD performance of a receiver. Figure 8.31 shows one standard set-up. Two signal generators are used to provide the two different signals required for the IMD test. Each signal generator

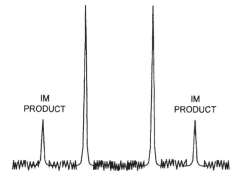

Figure 8.30
Third-order IM products are usually most important because they fall close to the test signals and may be within the receiver passband. Second-order IM products (F1 ± F2) tend to fall outside the band, so they are easily filtered out. Third-order IM products (2F1 ± F2 and 2F2 ± F1) are usually outside the band.

is equipped with an adjustable attenuator, which may or may not be external to the generator. In some cases, both internal and external attenuators may be used.

Optional bandpass filters are sometimes used to clean up the signal generator output spectrum. These filters are used to suppress harmonics of the output frequency. If the signal generator has sufficiently low harmonic output, then these filters can be eliminated. Keep in mind that some filters use ferrite or powdered iron cores, so may saturate and cause IMD products of their own. The two signals are combined in a two-port hybrid. Following the hybrid is another attenuator. This attenuator supplies signal to the receiver input.

The output signal is monitored by any of several means. Some procedures use the audio AC output level, as measured by an AC voltmeter. In other cases, the spectrum of the audio output signal is measured using a spectrum analyzer. Alternatively, one might also use a frequency selective voltmeter (also called a wavemeter). The latter method is out of favor because spectrum analyzer prices have fallen significantly. Some people will use the receiver S-meter (if it has one) to

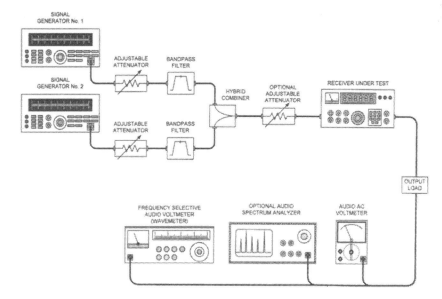

Figure 8.31 Test set-up for making complete measurement of IMD performance of the receiver.

make this measurement. Still others couple the IF signal to an RF/IF spectrum analyzer. The latter method may show more information, but it has the disadvantage of requiring entry inside the receiver. The other methods treat the receiver as a "black box," and thus require no modification of, or entry into, the receiver. The IMD test is best run in one of the linear reception modes (SSB or CW), but that is not always possible (e.g., when the receiver is FM-only or AM-only).

Audio Signal Level Method

The audio output level is monitored on an audio spectrum analyzer (or measured on a wavemeter). The signal levels are turned up until the IMD product being investigated rises up out of the noise level.

The spectrum analyzer method can be particularly useful for measuring products below the noise floor of the receiver. Recall that the noise floor is proportional to bandwidth. Typical bandwidths vary from 500 Hz for CW receivers to 200 kHz for FM broadcast receivers (and more for microwave radar receivers). If the bandwidth filter on the audio spectrum analyzer is set to some narrow value, such as 5 or 10 Hz, then the noise is much lower, so low-level IMD problems show up better.

Signal-to-Noise Ratio Method

This approach to measuring IMD uses either an audio signal-to-noise ratio meter or a SINAD meter. The audio output is set to produce a 1-kHz signal for this method. Care must be exercised to prevent excess noise contribution from the signal generator output noise. This noise is indistinguishable from receiver noise, so it makes the IMD look worse. It is also possible that AGC action will interfere with this test.

S-Meter Method

In this method the level of the IMD product is noted on the receiver's S-meter. A reference signal is then provided that matches the S-meter reading. This yields the level of the IMD product. Problems with this method include the fact that some receivers compress gain when the signal level gets to a level above S9 or S9 + 10 dB. This method is better than some, however, for measuring the IMD performance of receivers with very high IMD performance.

Standard Method

The normal method for measuring the IMD performance is to set the signal generators to some convenient high-level output (e.g., −20 dBm). Select test frequencies (F1 and F2) and calculate the third-order products (2F1 + F2, 2F1 − F2, 2F2 + F1, and 2F2 − F1).

Set the receiver to a channel frequency, F1. If possible, turn off the AGC or clamp it to a low value (highest receiver gain), if possible. If the receiver uses a front-end RF or IF attenuator, then set it to 0 dB. If there is an RF preamplifier being used, turn if off. Adjust both the receiver tuning and F1 to the same frequency, and maximize the receiver output. Set the second signal generator to F2 at a specified spacing (e.g., 20 kHz) away from F1. Set both signal generators to a convenient output level such as −10 dBm. Set the in-line attenuator to the highest setting (most attenuation).

Once the set-up is completed, turn off the signal generators and measure the receiver output noise level on an AC audio voltmeter. Turn on the signal generator

F1 and decrease the attenuator setting in 1-dB steps until the output noise level of the receiver increases 3 dB. This is the minimum discernible signal (MDS) reference level. Return the attenuator settings to maximum. Record this signal level as $P_{IM} = -10$ dBm $-$ (Attenuator setting).

Tune the receiver to either of the close-in third-order product frequencies (either 2F1 $-$ F2 or 2F2 $-$ F1), while leaving the signal generators at F1 and F2 (both -10 dBm output). Reduce the attenuator setting until the receiver output response at this frequency increases 3 dB (the same as the reference MDS). Record this level as $P_A = -10$ dBm $-$ (Attenuator setting).

$$IP_N = \frac{N\,P_A - P_{IM_N}}{N - 1} \qquad\qquad [8.14]$$

Where:

> IP_N is the intermodulation product of order N
>
> N is the order of the intermodulation product
>
> P_A and P_{IM} are signal power levels in dBm

EXAMPLE

A 162.55 MHz receiver was tested using frequencies F1 = 162.55 MHz and F2 = 162.57 MHz. The close-in third-order products would be 2F1 $-$ F2 = 162.53 MHz and 2F2 $-$ F2 = 162.59 MHz.

The minimum discernible signal at 162.55 MHz required an attenuator setting of -89 dB, so $P_{IM} = [-10$ dBM $- 89$ dB] $= -99$ dBm.

The response at 162.53 MHz required 19 dB of attenuation for the third-order response to equal the MDS level. So $P_A = [-10$ dBm $- 19$ dB] $= -29$ dBm.

Because this is a third-order response, N = 3, so

$$IP_3 = \frac{(3 \times (-29\ dBm)) - (-99\ dBm)}{3 - 1}$$

$$IP_3 = \frac{-87\ dBm + 99\ dBm}{2} = \frac{+12\ dBm}{2} = +6\ dBm$$

Once the P_A and P_{IM} points are found, any IP can be calculated using Equation 8.14.

Selectivity Testing

Selectivity is a measure of the receiver's ability to reject off-channel signals. It is measured in terms of bandpass, so it has the units of frequency. The operational definition of selectivity is that it is the bandpass (Figure 8.32A) between the two points (F1 and F2) where the frequency response drops -3 dB from its mid-band point (F_o).

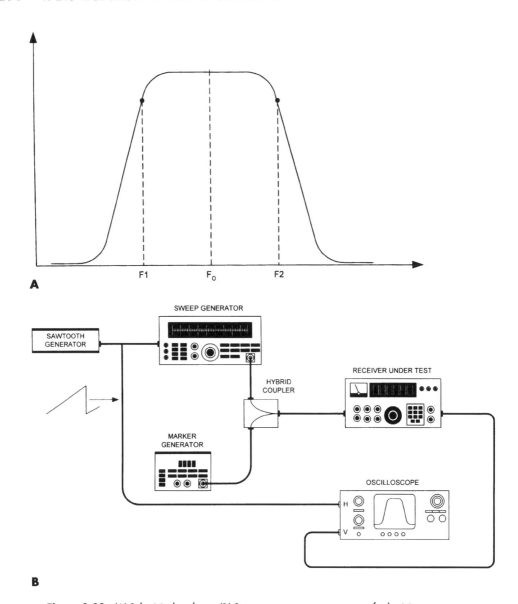

Figure 8.32 (A) Selectivity bandpass; (B) Sweep-generator measurement of selectivity.

CW Method

It is possible to use a standard CW generator to make this measurement, although it is tedious and may not yield the best information. Set up a signal generator and output indicator per Figure 8.24. Center the signal-generator frequency inside the receiver passband, and adjust the receiver output for a convenient level indicated on the AC voltmeter. Adjust the signal generator below F_0 to a point where the output drops the specified amount. This frequency is F1. Repeat the procedure above F_0 until F2 is determined. The receiver bandwidth is F2-F1.

Sweep Method

The CW method produces a raw indication, but lacks certain information, at least in its simplest form. We are not simply interested in the F2-F1 value, but also in the shape of the passband. Ideally, a receiver passband is flat inside the passband, rolling off gently on the upper and lower ends. That may not be the actual case. By using the sweep method, we can determine the shape of the passband, as well as its bandwidth. There are sometimes some surprises lurking there, so the information is needed. To get shape data by the CW method requires collecting a lot of data points and then manually graphing them. The sweep method produces an oscilloscope trace that can be photographed.

Figure 8.32B shows the test equipment set-up for a simple sweep test. A sweep generator will sweep through a specified band of frequencies. In the simplest form, it consists of a *voltage controlled oscillator* (VCO) with its tuning voltage in the form of a sawtooth waveform (modern sweepers are a bit more sophisticated, but the idea is the same). The same sawtooth can be used to control the horizontal deflection of an oscilloscope. The sweep signal is applied to the receiver's antenna input, while the receiver output is applied to the vertical deflection input of the oscilloscope. The result is a trace representing amplitude-versus-frequency.

A *marker generator* can be used to indicate specific frequencies within the passband (typically, F1, F2, and F_o). In some cases, discrete frequencies are used, while in others the harmonics of a standard frequency may be used. For example, a 1-kHz marker generator that is sufficiently rich in harmonics will provide markers every 1 kHz throughout the passband. There is, however, a limit to the use of harmonic markers, as their upper harmonics may not be strong enough to produce the desired display.

The sweep rate setting deserves some comment. The sweep rate is the repetition rate at which the signal passes through the swept band. This attribute is controlled by the frequency of the sawtooth generator. If the sawtooth rate is too slow, then the display will flicker and be hard to read. If it is too high, then there may be ringing in the receiver's IF filters, causing a distorted reading. Flicker fusion tends to occur in the 8- to 10-Hz range for most human operators, so the minimum sweep frequency will have to be at least these values. When you get up to above 40 Hz or so, the danger of causing a ringing response in high-Q IF filters becomes more likely. You should, therefore, select a sweep frequency between roughly 8 and 40 Hz for manually operated measurement systems.

Squelch Tests

Communications receivers are often equipped with a *squelch* circuit. These circuits turn off the receiver's audio output when no signal is being received. Noise occupies the bandwidth when no signal is received, and it is uncomfortable to hear. When the squelch circuit detects noise, rather than signal, it turns off the receiver output. The operator sets the squelch from the front panel.

There are two levels of squelch: *critical squelch* and *tight squelch*. The critical squelch level exists when the control is set so that it just barely quiets the receiver when no signal is being received. Indeed, when a particularly high-level noise burst is heard, it might degrade the squelch enough to pass through momentarily.

Tight squelch requires a much larger signal to cause break through, and it occurs when the squelch control is set to maximum.

Critical Squelch

Connect a signal generator to the antenna input of the receiver. Adjust the signal generator frequency on-channel.

Modulation. For AM receivers set the signal-generator modulation depth to 30% with a modulating frequency of 1,000 Hz. For FM receivers, use a deviation that is approximately 60% of the normal deviation for that receiver, with a 1,000-Hz modulating frequency.

1. Set the signal-generator output level to the lowest possible level so that no signal is heard in the receiver output.
2. Set the squelch control to the point where the output noise just disappears when no signal is present.
3. Bring the signal generator's output level up very slowly until the squelch breaks. Record the output level as the critical squelch point (dBm or μV).

Tight Squelch

The tight squelch test is performed identically to the critical squelch, but with the squelch control on the receiver set to the maximum (most squelched) position. The signal level required to break tight squelch will be much larger.

Squelch Range

This characteristic is the range between critical squelch and tight squelch. It may be expressed in signal level units (dBm or μV), or in dB. For example, suppose the signal levels are:

Critical Squelch: 0.4 μV (−115 dBm)
Tight Squelch: 58 μV (−71 dBm)
Squelch Range: 57.6 μV (44 dBm)

CHAPTER NINE

Radio Transmitter Measurements

The purpose of a radio transmitter is to generate a radio frequency (RF) signal and to impart information to it so that it can be propagated as a radio wave from an antenna. Although there are a number of variations on the architecture of the radio transmitter, Figure 9.1 shows the basic scheme. This particular architecture pertains to amplitude modulated (AM) transmitters, but the principal stages are found in others as well. The general term for this type of transmitter architecture is master oscillator power amplifier (MOPA).

The three functions of these stages are generating an RF signal, boosting it to a power level that is useful, and modulating it. The first of these functions, generating the signal, is found in the master oscillator.

Depending on the design of the transmitter, the master oscillator might be an L-C controlled variable frequency oscillator (VFO), a fixed-tuned crystal oscillator, or a frequency synthesizer. The VFO approach is no longer widely used; in the past (after about 1970), it had been used largely in amateur radio equipment. Fixed-tuned crystal oscillators are still very much in evidence. Some models use electronic or mechanical switching to select crystals for different frequencies. Crystal oscillators are found in room-temperature, temperature-compensated, and oven-controlled versions, depending on the accuracy and stability requirements imposed on the transmitter.

Increasingly, the frequency synthesizer is the MO design of choice, because it combines the stability and accuracy of the crystal oscillator with the flexibility of the VFO. Even single-channel transmitters and receivers use frequency-synthesizer master oscillators, because channel selection is reduced to applying the correct binary code to the divide-by-N counter in the synthesizer feedback loop. At one time, the use of frequency synthesizers would have been too costly in single-channel transmitters, but today the circuitry is available as a single-integrated circuit, and it is usually less costly overall to use the synthesizer approach.

The other required stage is the power amplifier (PA). The PA serves to boost the oscillator output to a level that makes it useful. The PA might be a 100-milliwatt amplifier used on short-distance telemetry transmitters, or a multimegawatt affair

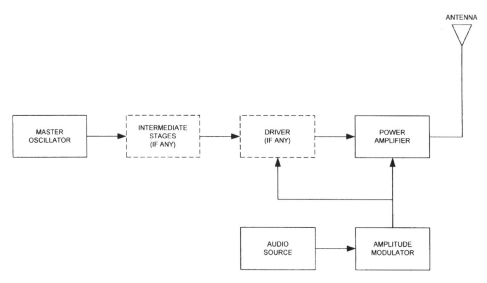

Figure 9.1 AM transmitter block diagram.

that is used for international broadcasting or tropospheric-scatter communications. Most communications transmitters fit somewhere between these extremes.

The power-amplifier stage requires some means of being coupled to the antenna, so it might have either a single-frequency resonant-coupling network, or a wideband impedance-matching network, in the output circuit. Maximum power transfer occurs between the PA and the antenna only when the source (PA) and load (antenna) impedances are matched.

The driver stage is used to provide an intermediate boost to the low-level signal produced by the MO stage. It is typically a power amplifier rather than a voltage amplifier, but it has a considerably lower power level than the power amplifier itself.

The driver and power amplifier also serve to isolate the MO from the antenna load. Oscillator circuits like to see a constant environment and are at their best when they produce low power levels. Almost all oscillator stages used in transmitters include a buffer amplifier between the oscillator output and the driver or power amplifier input. The buffer amplifier may be either unity or low gain, but has the advantage that it lightly loads the oscillator circuit, permitting it to run in a constant environment.

There may be other intermediate stages other than the buffer amplifier. For example, there may be frequency translation stages that use heterodyne mixing to change the frequency of the oscillator signal to some other frequency. There might also be frequency multipliers in use. The frequency multiplier approach is used in VHF/UHF transmitters where generating a crystal controlled signal is difficult. An oscillator will generate a signal in the 1- to 70-MHz region and then feed it to the multipliers. For example, suppose a transmitter has three multiplier stages: a doubler and two triplers. The frequency multiplication factor is therefore $2 \times 3 \times 3 = 18$. Suppose further that the transmitter has to operate on an assigned frequency of 408.75 MHz. The required frequency for the crystal is $408.75/18 = 22.7083$ MHz.

All frequency multipliers are nonlinear, thus offering a potential for producing a dirty output spectrum. Some transmitter maintenance instructions require the use of a spectrum analyzer to tune multiplier circuits in order to ensure that the correct harmonic is selected.

The MOPA architecture is more or less straightforward and represents many different types of transmitter. The way that transmitters differ with one another is in the way that information is imparted to the RF signal (called a "carrier"). In early transmitters, the carrier was turned on and off with a telegraph key to produce Morse code dots and dashes. Later, voice was added to the signal through the process of amplitude modulation. In Figure 9.1 we see the modulator stage and the preceding audio source (e.g., a preamplifier and a microphone).

Figure 9.2 shows the amplitude modulation (AM) process. The photo in Figure 9.2A shows the RF carrier signal (Fc), while Figure 9.2B shows the audio modulation signal (FM). When the two are combined, they produce the amplitude-modulated signal of Figure 9.2C. Note that the peaks of the RF carrier signal rise and fall in step with the applied modulating signal.

The process is multiplicative and nonlinear:

$$s(t) = A_c \ Sin \ (2\pi f_c \ t) \times [1 + \mu \ Sin(2\pi f_m \ t)] \qquad [9.1]$$

Where:

μ is the depth of modulation

f_m is the modulating frequency in Hertz (Hz)

f_c is the carrier frequency in Hertz (Hz)

t is time in seconds

Because the process is nonlinear it will produce a spectrum that includes the sum and difference frequencies, $F_C \pm F_M$. The spectrum for a radio carrier modulated with a single sine wave tone is shown in Figure 9.3. The sum component is called the upper sideband (USB), while the difference component is the lower sideband (LSB). It is interesting to note that the entire information content is inherent in either sideband, a fact that will prove interesting when we cover single-sideband transmitters. In the case of speech or other complex audio waveforms, the LSB and USB will spread out showing all of the sideband frequencies.

The way the RF carrier is modulated is a function of the design of the modulator. High-level modulation applies the modulation to the final amplifier, while low-level modulation applies it to the driver or an intermediate stage. One implication of the low-level design is that all stages following the modulated stage must be linear. A linear stage is less efficient than a non-linear (e.g., Class-C) amplifier, so there is a trade-off to be made.

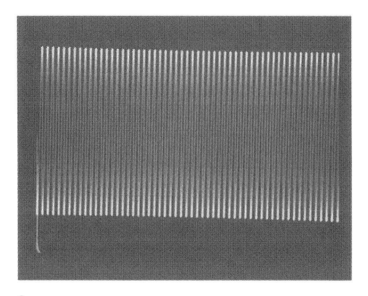

A

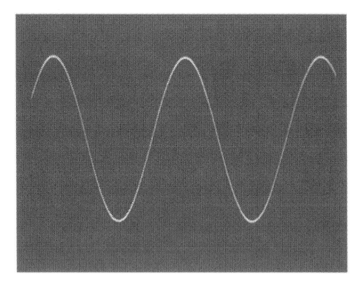

B

Figure 9.2   Amplitude modulation: (A) RF carrier; (B) Audio modulating signal; (C) AM modulated carrier.

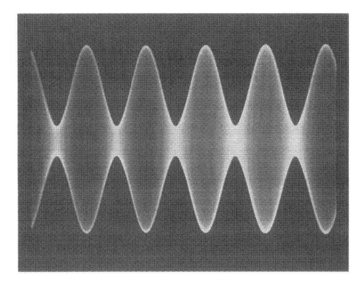

c

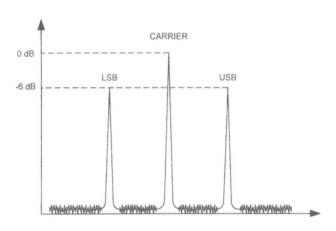

Figure 9.3
Spectrum created by sine wave modulation of a sine wave carrier. Note that two additional signals appear with the carrier, i.e., the upper and lower sidebands (USB and LSB).

PERCENTAGE OF MODULATION

Figure 9.4 shows both modulated and unmodulated segments of the RF signal. The RF carrier is a sine wave at an RF frequency. The unmodulated RF carrier voltage peaks at +E and −E. The audio signal is superimposed on it, producing swings from +EMIN to +EMAX on the positive side of the carrier sine wave, and −EMIN to +EMAX on the negative side of the carrier sine wave. The factors used to determine depth of modulation, also called percentage of modulation, are:

$$A = (+E_{MAX}) - (-E_{MAX})$$ [9.2]

$$B = (+E_{MIN}) - (-E_{MIN})$$ [9.3]

$$Percentage\ of\ Modulation = \frac{A - B}{A + B} \times 100\%$$ [9.4]

When the transmitter is fully (100%) modulated, the carrier is 6 dB higher than the sidebands. Figure 9.5 shows several interesting facts about AM that pertain to measurement. The percentage of the total RF power in the sidebands varies from zero when the percentage of modulation is zero (i.e., when it is unmodulated), to only 33% when the transmitter is 100% modulated. Given that there are two sidebands, and that all of the imparted information is carried in each sideband, that means only 16.5% of the RF power is used to transmit intelligence.

If you monitor the RF line current that is fed to the load during modulation, you will see it increase under modulation (Figure 9.5B). The percentage of change of the RF line current varies from zero at zero modulation, to 22.5% at 100% modulation.

Note in Figure 9.4 that the peak of the carrier varies with the modulating signal. This means that power increases and decreases with modulation. The percentage of modulation affects the ratio of this peak envelope power (PEP) according to the graph in Figure 9.5C. At 100% modulation, the PEP power is four times the unmodulated power level.

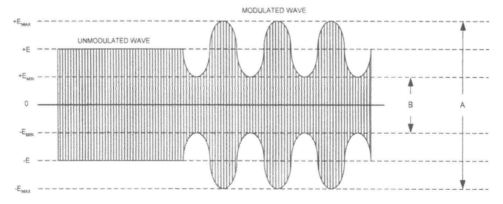

Figure 9.4 Standard relationships on a sine wave modulated RF carrier.

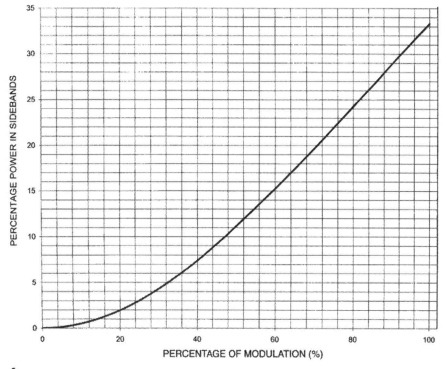

A

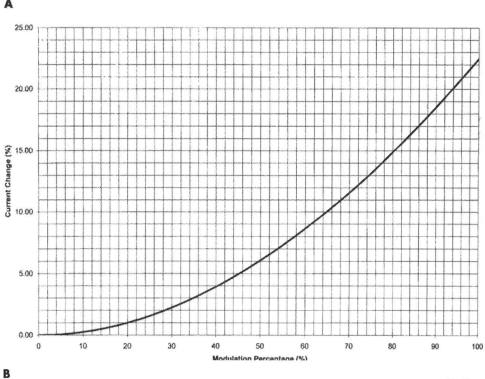

B

Continued

Figure 9.5 (A) Percentage of power in sidebands vs. percentage modulation; (B) RF current change vs. modulation percentage; (C) Ratio of peak envelope power (PEP) to unmodulated power vs. modulation percentage.

Figure 9.5 *(continued)*

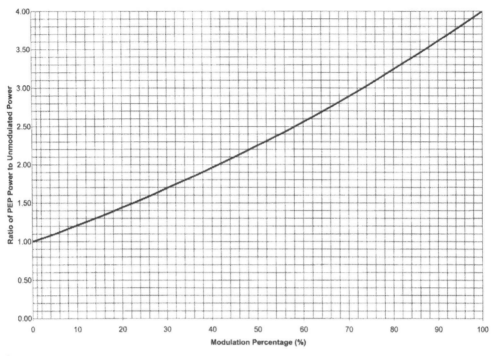

c

MEASURING RF POWER OUTPUT

The RF power can be measured with the set-up shown in Figure 9.6. In this instance, an RF wattmeter (see Chapter 6, "Radio Frequency Power Measurements") is connected to a dummy load. When the transmitter is keyed, the RF power is indicated on the wattmeter. Unless the wattmeter is designed for modulated signals, the unmodulated RF output power is used.

Note in Figure 9.6 that there is an adjustable DC power supply connected to the transmitter. This connection pertains to mobile and portable transmitters. Federal Communications Commission (FCC) regulations, nongovernmental standards of various types, and many manufacturer instruction manuals specify the supply voltage (e.g., 13.2 VDC) at which the RF power reading is to be made.

RF power can also be measured by measuring the RF voltage across the load ($P = E^2/R$) or the current flowing in the load ($P = I^2R$).

Dummy Loads

What is a *dummy load?* A dummy load is a *nonradiating substitute for an antenna.* That is, perhaps, why the British have traditionally called these devices "artificial aerials."

So why use a dummy load instead of radiating directly from the antenna? For several reasons. First, it is illegal in most countries to radiate a signal when testing

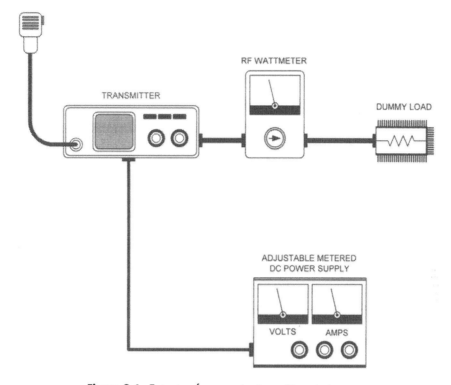

Figure 9.6   Test set-up for measuring transmitter output power.

transmitters if it interferes with other users on the channel. One is allowed to radiate only RF energy needed for communications. Another reason is that it is just plain rude to cause interference on a radio channel just because you want to test your transmitter. Rather than pressing the push-to-talk and interfering with other stations, we can silently key the power into a dummy load.

Finally, there is a very good technical reason for using a dummy load to test a transmitter: antennas cannot be relied upon to provide the constant and consistent test load that is necessary to make sense out of transmitter tests and adjustments. The measurements that you make may not match the specifications given in the transmitter's manual, even though there is nothing wrong. Figure 9.7 shows the standard circuit for the dummy load. It consists of a noninductive resistor mounted inside of a shielded enclosure, with either a coaxial connector or other transmission line connector to the outside world.

The resistor has to be noninductive so that the impedance it represents is similar to what would be seen on a resonant antenna. For most applications, an

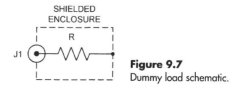

Figure 9.7
Dummy load schematic.

impedance of 50 ohms is used as the system impedance, although examples of 75-ohms, 300-ohms, 450-ohms, and 600-ohms systems are also occasionally seen. Most modern transmitters are designed to work into a 50-ohms resistive load.

Resistive Loads?

The impedance of any load can be described by:

$$Z = \sqrt{R^2 + (X_L - X_C)^2} \qquad\qquad [9.5]$$

Where:

R is the resistive component
X_L is the inductive reactance component
X_C is the capacitive reactance component

If an antenna is resonant, then $X_C = X_L$; the reactances thus cancel out, leaving only the resistive component. But not all antennas work directly on resonance, especially if they are required to work over a band of frequencies. When the transmitter frequency is lower than the resonant frequency, the antenna appears too short and exhibits some capacitive reactance (X_C). The usual solution is to add some inductive reactance (X_L) to cancel it out. Similarly, when the exciting frequency is above the resonant frequency, the antenna appears too long and exhibits inductive reactance. In both cases the impedance will not be resistive but complex. For that reason, an alternate form equation is sometimes used:

$$Z = R \pm jX \qquad\qquad [9.6]$$

The assumption that we will see only resistive loads is reasonable for some transmitters, but for others it is a fallacy. Whether or not it is true depends on the nature of the antenna system connected to the transmitter, and whether or not the transmitter stays on one frequency. For now, however, we will make the resistive assumption.

Simple Dummy Loads

There are a number of dummy loads that you can buy from commercial sources, but for now let's take a look at some basic forms. Figure 9.8A shows a simple low-power dummy load that can be used on HF QRP rigs, or many VHF transmitters up to the VHF bands. It consists of two or more parallel resistors (R) connected across either a male BNC connector or a PL-259 "UHF" coaxial connector, depending on the particular transmitter it is used to test.

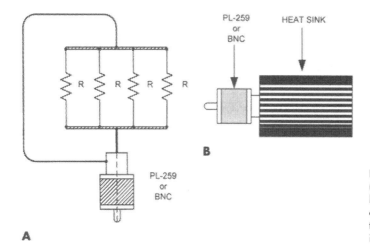

Figure 9.8
(A) Practical low-power dummy load; (B) Commercial examples often have heat sinks to increase the power level by dissipating heat in resistors.

The values of the resistors depend on the power level and the particular impedance being created. Let's assume 50 ohms for the overall impedance. If you place four 200-ohms resistors in parallel, then you will have a 50-ohms impedance. The power rating will be the total power rating of the resistors. For example, if you use 1-watt resistors, then it will be a 4-watt dummy load. Similarly, if the dummy load is made from 2-watt resistors, an 8-watt load is created. The resistors should be either carbon composition or metallic film. In no case should they be wire-wound resistors.

Higher power levels can be accommodated by using a larger number of resistors, with correspondingly higher value resistances in Figure 9.8A. Twenty 1,000-ohms, 2-watt resistors can be used to make a 40-watt load. However, keep in mind that the higher the number of resistors the greater the distributed capacitances, which essentially limits the frequency response of the load.

Small dummy loads are built in this manner. Figure 9.8B shows a representation of a common 20- to 50-watt dummy load used in both commercial and amateur communications testing.

Oil-Filled Paint-Can Dummy Antennas

A popular approach to making high-power ham radio (or impromptu commercial) dummy loads is the paint-can antenna. These loads were popularized by the old Heathkit *Cantenna*. Such a load consists of a standard one-gallon paint can. A noninductive 50-ohms power resistor is placed inside the can, and the can is then filled with either motor oil or mineral oil (at least one commercial model used a silicone oil, I am told). When searching for resistors, you must find noninductive resistors. Nearly all of the power resistors that you will find are wire-wound and not usable for a dummy load. You can recognize the noninductive resistor because it will be a ceramic cylinder with a coating of carbon-like material on the outer surface. In some cases, there will not be any connectors. The electrical connection is provided by hose clamps connected to the ends.

Air-Cooled Dummy Loads

The air-cooled dummy load is probably a lot more practical than oil-filled loads. The down side is that the power rating of the resistor element must be higher. Oil-filled loads can be operated at higher than rated powers because the oil couples heat up to the surface of the can where it can be radiated to the air around it. Commercial dummy loads at high powers (up to 50 kW are easily obtained) use either internal oil or an external water jacket to carry heat away from the resistor element.

Figure 9.9 shows a basic air-cooled dummy load. This figure represents a large number of 250-watt to 2,500-watt dummy loads. Most are built without the fan, although I've seen many commercially available models that have blower fan cut-outs on one end of the perforated aluminum cabinet. If you want to increase the power rating of the dummy load, then add the fan and get rid of the heat.

Another approach is shown in Figure 9.10. This type of dummy load uses a finned heat sink to radiate away the heat dissipated by the dummy load resistor. The internal cavity of the dummy load might be filled with either oil or a silicone gel material that matches the thermal impedance of the resistor to that of the shielded enclosure wall.

> Caution Note: A lot of ham-rated dummy loads have a short-duty cycle, so they might be less useful for commercial work. Look at the specifications of any load you obtain to find out how long you can keep the transmitter keyed without damaging the resistor. Some of them have remarkably short duty cycles. One model I saw said "60-seconds off for 10-seconds on." That means a one-minute cooling-off period is needed every time you key the transmitter for 10 seconds. If a particular model looks a bit too small for the wattage rating printed in big letters, then look at the fine print to see the duty-cycle rating.

Commercial Examples

Two commercial forms of dummy load are shown in Figures 9.11 and 9.12. The air-cooled version shown in Figure 9.11 uses a finned heat sink to remove the heat dissipated in the load. It can handle up to 300 watts of RF power. The types shown in

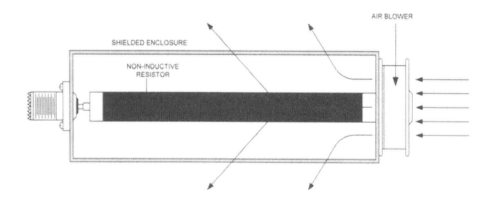

Figure 9.9 Medium- to high-power dummy load relies on air-cooling.

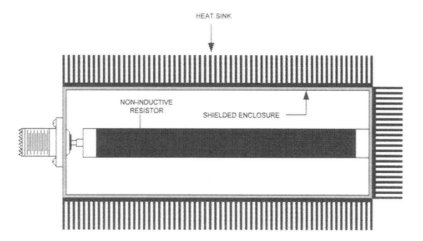

Figure 9.10 Medium- to high-power dummy load relies on heat sink to cool the resistor.

Figure 9.12 are rated at 10 kW to 40 kW. These models use flow-through water cooling to remove the heat dissipated in the load.

Complex Impedance Dummy Loads?

Virtually every commercial dummy load is a resistive impedance. But not all antennas are resonant at all frequencies within their bands of operation. Some have reactive components as well as resistive. In those instances suitable amounts of capacitive or inductive reactance will be needed.

RF SAMPLING DEVICES

Many of the tests described in this chapter require a low-power sample of the signal applied to the dummy load. There are several ways to provide this signal. Some dummy loads have sampling circuits built in, while in other cases an external device needs to be provided. Devices such as isolated hybrid combiners and directional couplers (Chapter 2, "Small Components Used in Radio Frequency Test and Measurement") can be used to provide a sample of the RF signal. Another approach is the coupled-tee connector shown in Figure 9.13. The IN/OUT path is a straight transmission line, to which the SAMPLED port is connected by a wire or small single-turn loop in close proximity to the transmission line. It picks off a small sample of the signal and feeds it to other instruments used in the measurement process.

Carrier Shift Meters

When the modulating sine wave signal is symmetrical—that is, it has no DC component—the modulation envelope of the transmitted signal is also symmetrical.

Figure 9.11   Bird Termaline™ dummy load. (Photo courtesy of Bird Electronics Corporation)

The condition called *carrier shift* occurs when a defect exists that unbalances the modulated signal. Figure 9.14A shows the basic circuit for making a carrier-shift measurement. The RF signal is sampled either using a pick-off loop (as shown in Figure 9.14A as L1) or a sampling tee as shown above, or one of the isolating couplers of Chapter 2. The sampled RF signal is demodulated with a diode detector and filtered to produce a DC level. Capacitor C2 has a value high enough to filter out fluctuations (e.g., 680 pF or so). The meter is adjusted to exactly midscale when the carrier is unmodu-

Figure 9.12 High power *Termaline*™ dummy loads. (Photo courtesy of Bird Electronics Corporation)

lated. If the carrier-shift condition exists, then the meter will move in a direction that indicates which peak has changed. The percent carrier shift is indicated by:

$$Carrier\ Shift\ (\%) = \frac{(I_C \pm I_M)\,(100\%)}{I_C} \qquad [9.7]$$

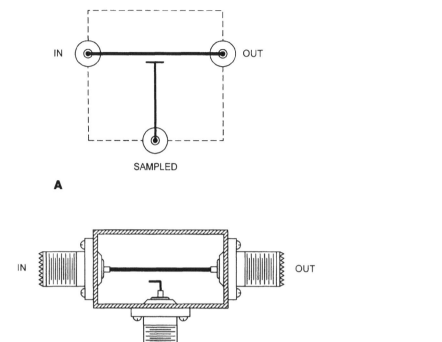

A

B

SAMPLED

Figure 9.13
Practical RF sampler: (A) schematic; (B) example of mechanical details.

Another approach is to view the signal on a spectrum analyzer that has the resolution to separate the sidebands and carrier (Figure 9.14B). In a normal AM signal the LSB and USB sidebands are of equal amplitude, but when carrier shift occurs, one of them will be of a lower amplitude than the other. The spectrum analyzer method is superior to the meter approach, but it is also more expensive. It is possible to use the spectrum analyzer to calibrate and validate the readings on the meter, but use the meter on a day-by-day basis.

PERCENTAGE OF MODULATION MEASUREMENTS (AM)

The measurement and testing of AM transmitters requires discovering the percentage of modulation and the degree of distortion present. The percentage of modulation is relatively easy to measure. One easy method is to display the modulated waveform on an oscilloscope, and then measure the various voltages seen in Figure 9.4. Any of the following relationships will yield the percentage of modulation:

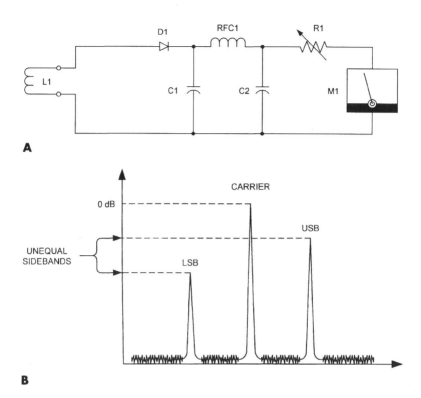

Figure 9.14 (A) Field-strength meter circuit can be adapted for making carrier shift measurements; (B) Unequal sidebands indicate carrier shift.

$$MOD\% = \frac{(E_{MAX} - E)\,(100\%)}{E} \qquad [9.8]$$

$$MOD\% = \frac{(E - E_{MIN})\,(100\%)}{E} \qquad [9.9]$$

$$MOD\% = \frac{(A - B)\,(100\%)}{(A + B)} \qquad [9.10]$$

These values can only be accurately read when the modulating signal is a sine wave.

One approach (Figure 9.15) is to use a commercial *modulation meter*. These instruments usually require a SET adjustment when the transmitter is unmodulated, followed by a measurement when the modulation is applied. Some modulation monitors are connected to the transmitter through a sampling tee as shown, while others have either a built-in dummy load or a pass-through path that allows the instrument to be inserted in the line between the transmitter and the load. The pass-through transmission line inside the instrument contains the sampling device. These instruments are often used with AM broadcasting and AM citizens band transmitters.

Envelope Detector Method

Most low-cost AM modulation meters are based on the envelope detector approach. An envelope detector circuit (Figure 9.16) is either connected to a sampling tee (in which case R_L is external) or is part of the dummy load R_L. Capacitor C1 has a value of a few picofarads in order to sample the signal without degrading the load impedance. It couples the signal to a diode voltage doubler that includes diodes D1 and D2, plus capacitor C2. The capacitor also serves to filter the output of the diodes to a DC level.

The output of the envelope detector is applied to the vertical input of an oscilloscope:

1. Set the input switch to GND (or "OV") to establish a ground baseline on the CRT grid (the line that is one division above the bottom-most line is usually a good start).
2. Set the 'scope for DC coupling, and key the transmitter without modulation. This establishes the voltage that represents the peak voltage of the unmodulated carrier (E_C). Record this voltage.
3. Modulate the transmitter. The 'scope trace will rise to 2× the E_C on positive peaks and zero on negative peaks. Measure the peak-to-peak values of the modulated waveform and calculate:

$$MOD\% = \frac{(E_{P-P})\,(100\%)}{2\,E_C}$$ [9.11]

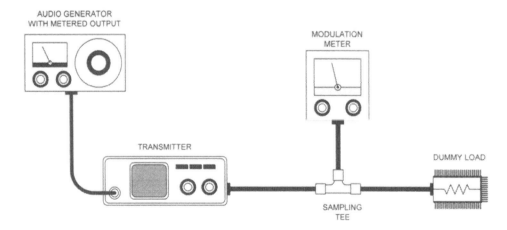

Figure 9.15 Using a modulation meter to measure AM percentage of modulation.

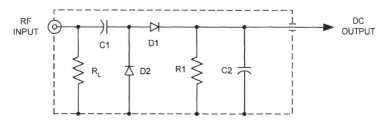

Figure 9.16 Envelope detector circuit.

Note: E_{P-P} is the voltage measured from the zero base to the top of the positive peak.

Trapezoid Method

The trapezoid method is second only to the spectrum analyzer method in evaluating amplitude modulated transmitters. It will reveal the percentage of modulation, as well as whether or not distortion is present. The trade-off between the trapezoid method and the spectrum analyzer method is one of convenience and cost. The trapezoid method can be performed with an ordinary oscilloscope, while a spectrum analyzer is an instrument that must be procured separately. However, the best measurement using the trapezoid method requires access to the output of the modulator.

Figure 9.17 shows the basic set-up for the trapezoid method. The modulated RF signal is applied to the horizontal input of the oscilloscope, while the modulating audio signal is applied to the vertical. If the modulating audio signal is taken from the output of the modulator stages, then the trapezoid is capable of showing distortion arising in the modulator stages.

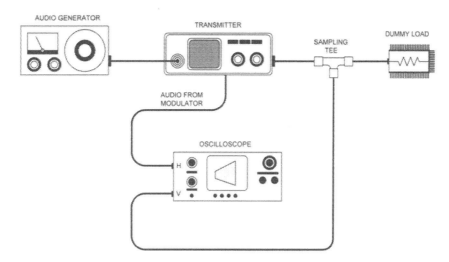

Figure 9.17 Measurement of AM percentage of modulation by the trapezoid method.

In mobile AM transceivers, such as citizens band, the audio output stages of the receiver are also used for the transmitter modulator. It is usually possible to find a point in that circuit for tapping the audio signal to the RF amplifiers.

If the high-level signal at the modulator output is not available, then one can use a demodulator probe (envelope detector) to recover it from the modulated RF signal. Alternatively, the worst case (but still moderately useful for percentage of modulation tests) is to use the input modulating audio signal. The added value of using the audio signal at the point where it is applied to the RF stage is, however, so significant that it should be done wherever possible. Figure 9.18 shows the relationship between the modulating audio sine wave and the vertical deflection caused by the RF signal. When these two signals combine on the CRT screen the result is the trapezoidal pattern shown. Patterns for both 100% (Figure 9.18A) modulation and less than 100% (Figure 9.18B) are shown for comparison. In the cases where the percentage of modulation is less than 100%, the actual percentage of modulation can be obtained by measuring the two vertical edges (Figure 9.19) and calculating:

$$\%MOD = \frac{(A - B)\,(100\%)}{A + B} \qquad [9.12]$$

The patterns shown in Figure 9.20 are for various stages of modulation:

1. No RF signal
2. RF signal unmodulated
3. Less-than-100%
4. 100%
5. Overmodulation (i.e., attempt to get more than 100% modulation). The last trace would indicate a transmitter capable of considerable interfering "splatter" to adjacent channels.

One of the advantages of the trapezoid method is that distortion is relatively easy to see. If you examine only the envelope of the modulated signal on an oscilloscope, you may see distortion as a flattening of the peaks. But considerable amounts of distortion can occur before the clipping becomes easily recognized. Using the trapezoid method overcomes that problem because the nonlinearities show up as curved edges on the pattern (Figure 9.21).

Various pathologies are shown in Figure 9.22. The pattern in Figure 9.22A is the same pattern as before and indicates that distortion is present. In Figure 9.22B, the pattern indicates that the modulator is not capable of delivering the power required to fully modulate the RF amplifier stages. In Figure 9.22C there is a phase shift indicated by the dual-trace of the trapezoid pattern. This could indicate a phase shift problem in the transmitter, or it could indicate some problem in the test and measurement set-up. Always be sure to check the set-up before assuming phase shift problems in the transmitter.

Distortion Measurements

Distortion comes in two flavors: total harmonic distortion (THD) and intermodulation distortion (IMD). Both types produce spurious products at frequencies away from the carrier. The difference is that these products are harmonics (integer mul-

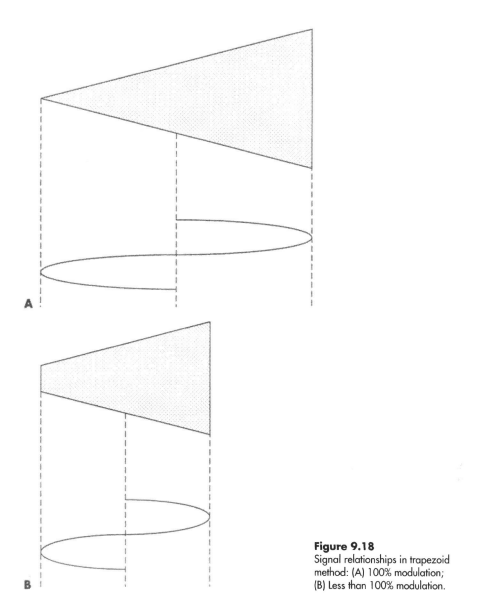

Figure 9.18
Signal relationships in trapezoid method: (A) 100% modulation; (B) Less than 100% modulation.

tiples) of the modulating frequency in THD, and in IMD they meet the mF1 ± nF2 rule. In this section we are concerned primarily with THD.

Single-Trace Method

If the distortion is profound then we can use the single-trace oscilloscope method. Observe the envelope waveform on an oscilloscope while the RF signal is modulated with a sine wave audio tone. The effect of THD is to distort the modulated waveshape, most frequently by flattening the tops of the peaks. The difference is that it is difficult to see small amounts of distortion using the single-trace method, even though they have an adverse effect on operation. A slightly better approach is the dual-trace method.

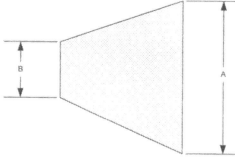

$$Percent\ Modulation\ =\ \frac{A - B}{A + B}\ \times\ 100\%$$

Figure 9.19
Measuring modulation percentage using trapezoid method uses these relationships.

Dual-Trace Method

The dual-trace method requires displaying the modulated RF envelope on one trace, and the modulating audio on the second, of a dual-channel oscilloscope. If done correctly, the two traces will overlay each other exactly. The position control is used to place the audio over the corresponding portions of the modulated RF signal. Departure from ideal, which indicates distortion, is seen by the displays that are not overlapping. This method is somewhat better than the single-trace method, but it still requires relatively large amounts of distortion before it is sensitive to the difference. Both methods suffer from the inability to provide numerical information on the degree of distortion.

Distortion Analyzer Method

A distortion analyzer or "distortion meter" is essentially an audio AC voltmeter with a switchable notch filter at the input. The AC voltage of the demodulated RF carrier (V1) is measured with the filter out of the circuit, providing the overall signal level. Next, the filter is switched into the circuit and tuned to notch out the fundamental. If the transmitter was modulated with a 1,000-Hz tone, then tune the notch filter to 1,000 Hz. The voltage is measured again (V2). This measurement represents the component due to harmonics. The Total Harmonic Distortion is:

$$THD\% = \frac{V2 \times 100\%}{V1} \qquad [9.13]$$

A harmonic distortion analyzer will contain the voltmeter and the filter. When intended specifically for off-the-air AM transmitter measurements, it will also include an envelope detector and filter.

Figure 9.23 shows the connection scheme for a THD analyzer that is not specifically designed for AM transmitter measurements. The input to the analyzer is fed from either an external envelope detector, or from a modulation meter that has its own internal envelope detector. The signal is derived from the sampling-tee junction of an isolated port-coupler device.

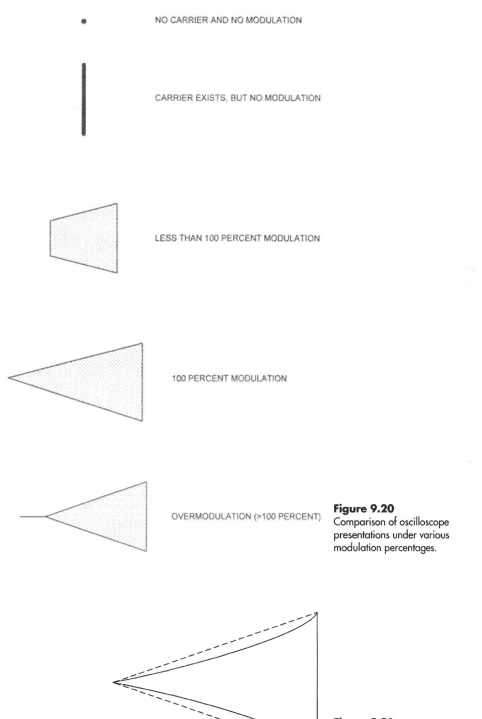

NO CARRIER AND NO MODULATION

CARRIER EXISTS, BUT NO MODULATION

LESS THAN 100 PERCENT MODULATION

100 PERCENT MODULATION

OVERMODULATION (>100 PERCENT)

Figure 9.20
Comparison of oscilloscope presentations under various modulation percentages.

Figure 9.21
Nonlinearity shows up in distortion of the sides of trapezoidal pattern.

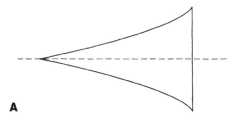

A

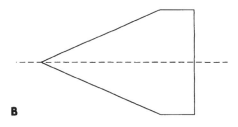

B

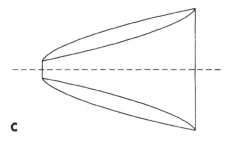

Figure 9.22
Various signal pathologies show up
on trapezoid.

C

Spectrum Analyzer Method

The spectrum analyzer represents the best approach to measuring distortion, because it will show both THD and IMD. Figure 9.24 shows the connection scheme. The RF input to the spectrum analyzer is taken from the sampling tee or coupler. If the signal is still too strong for the spectrum analyzer input, then an external attenuator can be inserted. In good practice, the attenuator is used in any event, simply to discern the difference between distortion products generated inside the spectrum analyzer and distortion products that are actually present in the signal.

Figure 9.25A shows the spectrum expected when a single sine wave tone is used to modulate the AM transmitter. The LSB and USB signals are separated from the carrier by a frequency difference equal to the frequency of the modulating au-

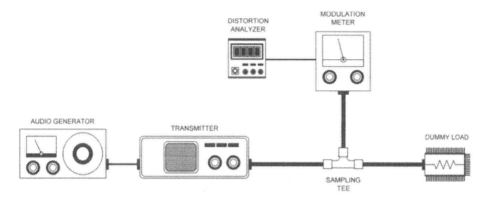

Figure 9.23 Measuring total harmonic distortion of AM signal using a distortion analyzer.

dio. If there is distortion present, additional products are generated to form a pattern similar to the one shown in Figure 9.25B. The difference between THD and IMD is the specific frequency products that are generated.

Single Sideband (SSB) Transmitters

Single sideband (SSB) is a species of amplitude modulation in which the carrier and one sideband are suppressed. Figure 9.26 shows this effect. The upper sideband (USB) case is shown in Figure 9.26A. Note that the lower sideband and carrier are suppressed, leaving only the upper sideband. The opposite case is shown in Figure 9.26B. This is the lower sideband (LSB) case in which the carrier and upper sideband are suppressed, leaving only the LSB.

Why would one want to go to the trouble of suppressing the carrier and one sideband? Recall from our discussion of AM, earlier in this chapter, that all of the information being transmitted is carried in either sideband. Thus, the carrier and one sideband are basically along for a free ride. This reduces the amount of

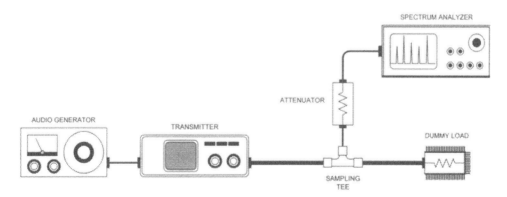

Figure 9.24 Measuring total harmonic distortion of AM signal using a spectrum analyzer.

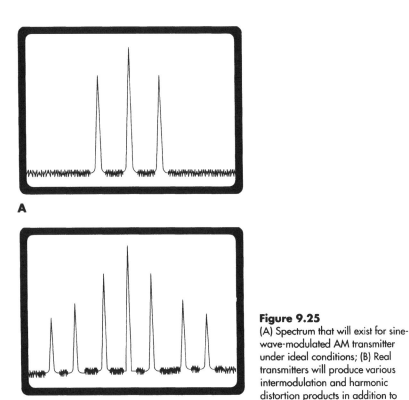

Figure 9.25
(A) Spectrum that will exist for sine-wave-modulated AM transmitter under ideal conditions; (B) Real transmitters will produce various intermodulation and harmonic distortion products in addition to LSB and USB.

the total RF power that is effective. The SSB transmitter is able to use all of its power for information transmission, so it has an advantage over the straight AM transmitter.

When the receiver picks up the SSB signal, it must provide a locally generated substitute carrier in order to demodulate the signal. SSB receivers therefore use product detectors rather than envelope detectors. The product detector is an RF mixer that heterodynes the locally generated carrier (from a beat-frequency oscillator) against the SSB signal at the output of the IF amplifier, to recover the audio difference frequencies.

The SSB Transmitter

There are two basic ways to generate the SSB signal: phasing and filter. Of these two, the filter method is by far the most common, so let's consider it. The block diagram of a filter-style SSB transmitter is shown in Figure 9.27.

Carrier suppression is accomplished first by combining an RF carrier-oscillator (F_C) signal with the modulating audio signal (F_A). The double-balanced modulator has the attribute of producing the standard mixer spectrum ($mF_C \pm nF_A$), while suppressing the two input frequencies (F_C and F_A). Thus, when used in a DSB generator, it will suppress the first-order signals, that is, the carrier signal and

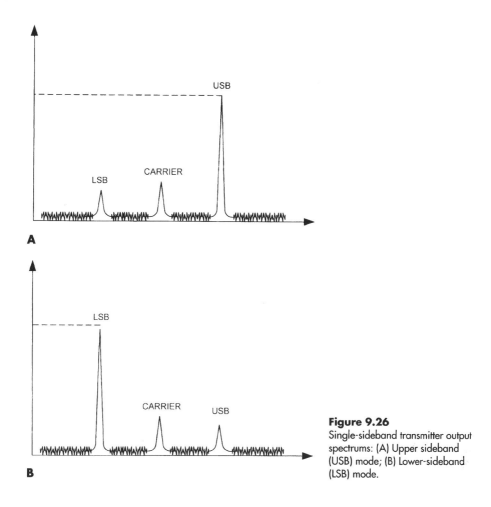

Figure 9.26
Single-sideband transmitter output spectrums: (A) Upper sideband (USB) mode; (B) Lower-sideband (LSB) mode.

the modulating audio signal. The output spectrum will contain the sum and difference products $F_C + F_A$ (USB) and $F_C - F_A$ (LSB).

The sum and difference signals represent the USB and LSB, respectively. To suppress the unwanted sideband, the signal is passed through a narrow RF filter (2.1 to 3 kHz BW, typically) that accepts the desired sideband, and rejects the unwanted sideband.

Some transmitters allow switching between USB and LSB. There are two approaches to this job. The first is to select two different filters, one centered above F_C and the other below. Although this approach is taken in some transmitters, it is rare. The cost of SSB-quality filters is high, so it is more common to take the second approach: that is, to use a single filter, but to have different carrier frequencies for LSB and USB. Figure 9.28 shows this arrangement: the carrier oscillator frequency is placed either immediately above or immediately below the filter skirt, depending on whether USB or LSB is desired.

It is rare to find a transmitter where the SSB signal can be generated on a single frequency that is transmitted over the air. In most designs, frequency flexibility

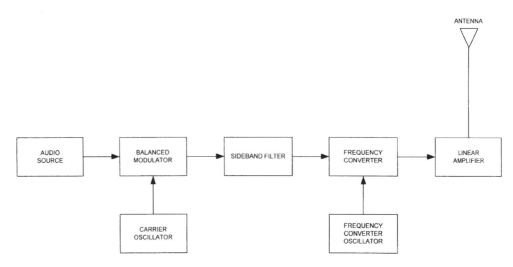

Figure 9.27 Filter-method for generating a single-sideband (SSB) signal.

is required. For example, in a marine H.F. SSB transmitter there are a number of H.F. voice bands. Each of these would require its own SSB generator. However, if some intermediate frequency is selected, a high-quality SSB signal can be generated and then translated to the band of choice.

In Figure 9.27 the frequency-converter mixer and frequency-converter oscillator perform this function. An SSB signal will be generated at some intermediate frequency (e.g., 455 kHz, 8.83 MHz, and 9 MHz are popular choices), and then is heterodyned in the frequency translator to the frequency of choice. The converter oscillator might be a single-frequency crystal oscillator, a variable-frequency oscillator, or (increasingly the universal case) a frequency synthesizer.

In this era of microprocessor-controlled frequency synthesizers, an interesting design trend has been noted in SSB transmitters. The frequency synthesizer chip may have a set of one to three address pins that are grounded for a binary LOW (0), or left ungrounded for a binary HIGH (1). The particular code sensed at the turn-on tells the microprocessor which set of frequencies are authorized for this transmitter. For example, one manufacturer made three SSB transmitters that appeared identical on the inside. One was for land-fixed and mobile applications, one for maritime use, and the third for amateur radio use. Two pins on the synthesizer would tell the transmitter which set of frequency look-up tables were authorized:

00	All frequencies allowed
01	Maritime frequencies allowed
10	Land-fixed/mobile frequencies allowed
11	Amateur radio frequencies allowed

Interestingly enough, there was a tremendous price differential between the 01, 10, and 11 codes (the 00 code was not sold, to the best of my knowledge). It should not be surprising that some enterprising technicians figured out how to make one radio work on the other sets of frequencies. In at least one case in central

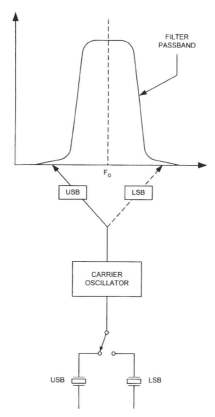

Figure 9.28
A single filter can be used to generate LSB and USB by changing the frequency of the carrier oscillator.

Africa, a technician of my acquaintance modified amateur-radio-coded equipment for his oil company employer to use as fixed/mobile land transmitters.

SSB POWER RELATIONSHIPS

The power relationships in an SSB signal can be complicated. The *peak envelope power* (PEP) is usually specified. When the modulating signal is a single sine wave tone, then the PEP is equivalent to the average power ($P_{PEP} = P_{AV}$), so an ordinary peak reading or average reading RF watt-meter can be used.

On complex waveforms, where the modulating signal contains more than one frequency, the relationships change quite a bit. The P_{PEP}/P_{AV} ratio changes dramatically as the number of frequencies present in the modulating signal increases. For the sake of measurement consistency (and ease, if truth be known) it is common to use a two-tone audio-test signal for measuring SSB transmitters.

The usual test signal for measuring the SSB transmitter consists of two equal amplitude sine wave tones that are *not harmonically related* (e.g., 500 Hz and 2,400 Hz). Figure 9.29 shows the voltage envelope displayed on an oscilloscope of a

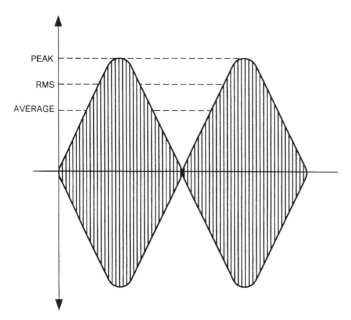

Figure 9.29
Voltage envelope displayed on an oscilloscope of a two-tone SSB RF signal.

two-tone SSB RF signal. The peak, RMS, and average voltages are shown. The *Instantaneous Peak Power* (IPP) is found from

$$IPP = \frac{V_P^2}{R_L} \qquad [9.14]$$

Where:

IPP is the instantaneous peak power (watts)
V_P is the peak voltage on the envelope
R_L is the load resistance

By way of example, let's assume that the standard 50-ohms load is used for R_L and that the peak voltage is 100 volts. The instantaneous peak power is:

$$IPP = \frac{(100V)^2}{50\ Ohms} = 200\ watts$$

In the two-tone case the IPP value is not the same as the PEP value because PEP is based on the RMS value of the two-tone waveform:

$$PEP = \frac{\left(\frac{V_P}{\sqrt{2}}\right)^2}{R_L}$$

Using the same case as above, the RMS value of the waveform is $100V/\sqrt{2} = 100V/1.414 = 70.712$ volts. The peak envelope power is:

$$PEP = \frac{(70.7V)^2}{50\ Ohms}$$

Note that for a two-tone test signal the value of the PEP power is one-half the IPP value.

To further complicate matters, the typical RF power meter is an *average-reading, RMS-calibrated* instrument. In that case, the meter reading would be about 40% of the PEP value. As a rule of thumb, when using average-reading, RMS-calibrated RF watt-meters, the *Watt-meter reading = 40.5% PEP = 81% of true P_{AV}*.

Two-Tone Test Set-up

The two-tone test is used for SSB transmitter testing because it more nearly represents the actual situation when complex modulating waveforms are applied, yet it retains much of the simplicity and ease of interpretation of single-tone testing. Figure 9.30 shows a typical test set-up. Select the two tones to be reasonably wide-spaced within the modulating spectrum (e.g., 300 to 3,000 Hz for voice communications), but they must not be harmonically related. Two popular tones are 500 Hz and 2,400-Hz. These tones come from separate audio oscillators and are combined into a single port by a linear combiner network of some sort. The oscillators chosen should be high-quality devices with little distortion and noise present.

The audio tone sources are not always two separate signal-generator instruments as shown, but may be a pair of audio oscillators in one circuit. There are special two-tone audio generators available specifically for two-tone SSB testing.

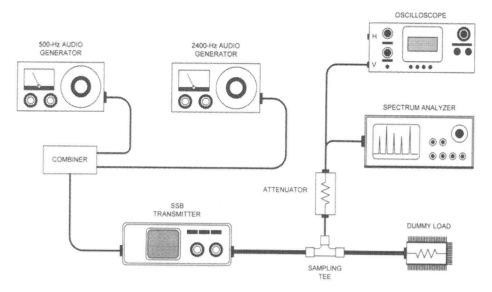

Figure 9.30 Two-tone SSB test set-up.

The output signal is sampled from the dummy load and transmission line and is applied at least to an oscilloscope. If available, a spectrum analyzer can also be used. Some measurements, such as carrier or unwanted sideband suppression, are actually better made on a spectrum analyzer.

There are several approaches to making the linear combiner network. It must be stressed that this is a linear mixer, so the use of active devices is usually discouraged. One approach is shown in Figure 9.31. A potentiometer is connected to the outputs of the two signal sources, with the combined signal taken from the potentiometer wiper terminal. Using a potentiometer instead of two fixed resistors also serves the function of providing a balance control. The two audio tones are to have equal amplitude, and the balance control is one means for smoothing out differences between the two signals.

The use of a T-pad linear network is shown in Figure 9.32. In this circuit, a number of 600-ohms resistors are used to blend the two tones together into a single output line. The 600-ohms value is selected to match the output impedance of standard audio signal generators.

Although it does not provide balance the way the potentiometer circuit does, it provides improved isolation between the two signal generators. The respective tone-signal levels are reduced at the output terminal by voltage-divider action, but the reduction is even more for the unwanted signal presented to the output circuit of the other generator. In other words, because all of the resistors are equal, the signal level of each oscillator is reduced by one voltage divider to the combined output port, but by two voltage dividers at the other oscillator's output.

The circuits of Figures 9.31 and 9.32 work well for unbalanced audio inputs. Some larger commercial SSB transmitters, however, use the same sort of standard 600-ohms balanced line commonly found in radio broadcasting transmitters. For SSB transmitters the circuit of Figure 9.33 will provide the two tones. Transformers T1 and T2 are standard 600-ohms line-to-line transformers and are available from most professional electronics parts catalogs.

Unwanted Carrier and Sideband Testing

The quality of an SSB signal depends on many of the same things as an AM signal: lack of hum, noise, and audio distortion. Additionally, however, the driving factors are the degree of suppression of the carrier and the unwanted sideband. No transmitter will completely suppress these signals, but the greater the suppression the

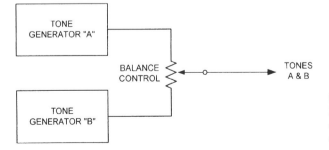

Figure 9.31
Linear combiner network makes it easy to balance the tone amplitudes.

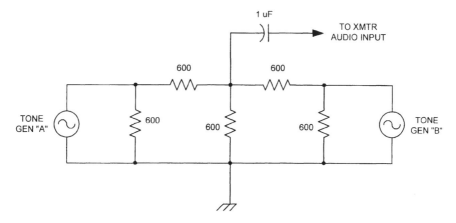

Figure 9.32 Linear combiner network for two-tone tests.

better the SSB signal. It would be really great if a transmitter could reduce the unwanted signals to just below the noise floor. But that's a dream. As a consequence, it becomes necessary to measure the degree of suppression.

The measurement of unwanted carrier suppression is one area where we will use a single-tone test set-up. The normal test set-up is exactly like Figure 9.30, but with one signal generator and the combiner deleted. Indeed, some SSB test sets that contain internal audio oscillators sometimes simply turn off the unwanted frequency.

When a single sine wave tone is used to modulate an SSB transmitter, the RF envelope will resemble a CW waveform, as shown in Figure 9.34. But if there is carrier or unwanted sideband signals present, then the result will look something like Figure 9.34B. In either case, the RF envelope peaks will take on a ripple, as shown. The ratio of the ripple to the overall envelope height is the measure of the suppression.

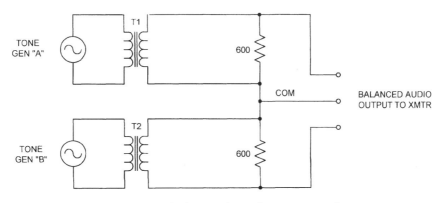

Figure 9.33 Hybrid approach to making two-tone combiner.

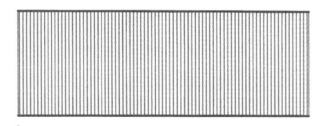

A

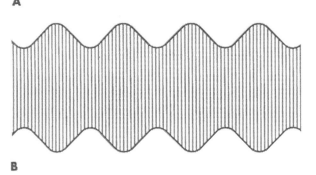

B

Figure 9.34
(A) When a single sine wave audio tone is used to modulate an SSB transmitter, the RF envelope will resemble a CW waveform;
(B) Carrier or unwanted sideband signals present on carrier.

Distinguishing the Carrier from the Sideband

The waveform of the envelope seen on an oscilloscope is nearly the same for both unwanted carrier and unwanted sideband. Discerning the difference requires comparing the frequency of the ripple with the frequency (or period) of the modulating audio signal. Consider the case where a 1,000-Hz tone is used to modulate the RF signal. The period of 1,000 Hz = 1/1000 Hz = 0.001 s = 1 ms.

Figure 9.35 compares both the poor carrier suppression and poor unwanted sideband suppression cases with the modulating audio. Here are two rules:

- Poor carrier suppression is indicated by a ripple frequency equal to the modulating audio frequency.
- Poor unwanted sideband suppression is indicated by a ripple frequency equal to twice the modulating audio frequency.

On an oscilloscope we would normally measure period and then convert it to frequency by taking the reciprocal. But since we know the period of the modulating signal is 1 ms, we can measure the ripple period and make the determination. If the ripple period is 1 ms on the 'scope, then the problem is poor carrier suppression. But if the period of the ripple component is 0.5 ms (i.e., 500 ns), then the problem is poor unwanted sideband suppression.

Measurement of Suppression

The degree of suppression of either the unwanted sideband or the carrier is expressed in decibels (dB) and is found by taking the ratio of the RF envelope ampli-

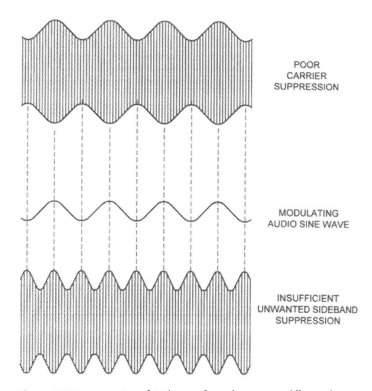

Figure 9.35 Comparison of similar waveforms demonstrates difference between poor carrier suppression (ripple = audio frequency) and insufficient unwanted sideband suppression (ripple = 2× audio frequency).

tude ("A" in Figure 9.36) and the ripple amplitude ("B" in Figure 9.36). The suppression is:

$$Suppression = 20 \, LOG \left[\frac{A}{B} \right] \qquad [9.15]$$

The same information can be obtained from a spectrum analyzer by comparing the heights of the carrier and/or unwanted sideband relative to the height of the desired sideband. If the spectrum analyzer screen is already calibrated in decibels (as some are), then the measurement is simply an observation. But if a linear scale is used the previous equation will yield the dB difference.

Testing SSB Transmitter Stages Using Trapezoid Method

A single-tone sine wave and an oscilloscope can be used to check the linearity of each stage in an SSB transmitter. Because SSB tends to be generated at a low level,

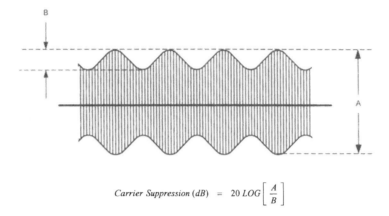

$$Carrier\ Suppression\ (dB)\ =\ 20\ LOG\left[\frac{A}{B}\right]$$

Figure 9.36 Measurement of carrier suppression in decibels.

all amplifier stages following the SSB generator must be linear. Figure 9.37 shows the test set-up. The basic method is the same sort of trapezoid method used for AM transmitters, except that the audio is derived by demodulating the RF signal at various stages within the transmitter (see inset to Figure 9.37). The RF signal can be obtained with a sampling tee, isolated coupler, or the resistive voltage divider shown as part of the dummy load in Figure 9.37. The percentage of modulation and the linearity are shown by the trapezoidal pattern in the same way as for AM transmitters.

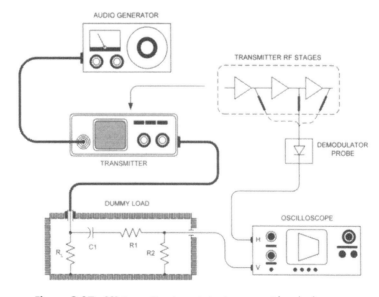

Figure 9.37 SSB transmitter stages test using trapezoid method set-up.

FREQUENCY MODULATION (FM) AND PHASE MODULATION (PM)

There are several different ways to impart information to a radio carrier signal. In this section we will examine two forms of *angular modulation: frequency modulation* (FM) and *phase modulation* (PM). These two are essentially equivalent, even though they are not the same; thus they will be discussed as one form, except for certain specific differences.

Among the reasons to use FM/PM is that it has the ability to provide a higher signal-to-noise ratio (SNR) than AM or SSB for a given RF power level. FM/PM transmissions can be more or less noise-free when demodulated on an appropriate receiver, even with relatively low signal levels. For this reason, high-fidelity broadcasting uses FM.

FM/PM is not used below about 30 MHz because it requires a relatively large bandwidth and is subject to sky-wave distortion when propagated in the ionosphere (AM is also so afflicted, but less so, especially when the LSB, carrier, and USB fade out of phase). Although SSB is also useful, the frequency tolerances required are much more stringent.

Figure 9.38A shows an unmodulated CW carrier. When no modulation is present, the FM/PM signal is essentially a CW signal as shown here. But when an audio signal is applied to the modulator, the frequency (in FM) or phase (in PM) begins to vary in step with the modulating audio (see Figure 9.38B).

Deviation and Swing

Figure 9.38B shows that the frequency varies from a low value (F_L) to a high value (F_H), above and below the frequency of the unmodulated carrier (F_O). The frequency difference between F_O and either F_L or F_H is called the *deviation* (F_D). Deviation is therefore measured in units of frequency. Notice that peak deviation on both sides of the carrier frequency corresponds to the peak amplitude of the modulating audio. The *frequency swing* is $F_H - F_L$ and will be twice the deviation.

One distinct difference between AM and either FM or PM is the nature of the sidebands. In an AM signal, when a carrier (F_C) is modulated by an audio signal (F_M), there will be a single set of sidebands generated equal to $F_C \pm F_M$. But in FM/PM there are a number of sidebands generated. The specific set of sidebands produced is a function of the deviation (F_D) and modulating frequency (F_M). The ratio of these frequencies is called the modulation index (β):

$$\beta = \frac{F_D}{F_M} \qquad\qquad [9.16]$$

Where:

β is the modulation index

F_D is the deviation

F_M is the modulating frequency

(Both F_M and F_D are in the same units)

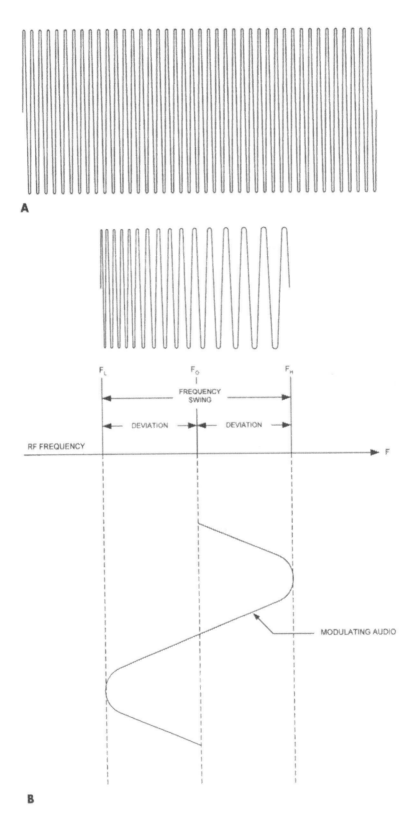

A

B

Figure 9.38 (A) Unmodulated RF carrier; (B) RF carrier frequency modulated by a sine wave tone.

For example, if an 800-Hz sine wave tone is used to modulate an FM transmitter to a deviation of 5 kHz, the modulation index is $5,000/800 = 6.25$. The modulation index also indicates the phase deviation in radians.

Another key parameter of the FM/PM transmitter is the *deviation ratio*. This parameter is the ratio of the peak deviation ($F_{D(peak)}$) to the highest modulating frequency ($F_{M(Max)}$):

$$\delta = \frac{F_{D\,(Peak)}}{F_{M\,(Max)}} \qquad\qquad [9.17]$$

The FM/PM signal is usually wider than an AM signal because of the sidebands generated. Figure 9.39 shows how the carrier and sidebands vary with modulation index. In due course we will take advantage of the fact that the carrier goes through zero (it disappears!) at certain values of modulation index.

Comparing an FM signal to an AM signal for bandwidth is not easily done. But for a single-tone modulating frequency, a modulation index of about 0.65 will correspond about equally to an AM signal. Single-tone performance, however, does not adequately predict the performance when a complex waveform such as speech is applied. There is a generalized rule relating the bandwidth requirements for a given FM/PM signal when narrow band FM is used:

$$BW = 2 \times (F_{M\,(MAX)} + F_{D\,(Peak)}) \qquad\qquad [9.18]$$

FM/PM Transmitters

Figure 9.40 shows the block diagram for an FM or PM transmitter. Note first the stages between the microphone and the audio amplifier. The first stage is labeled *pre-emphasis* and is used on FM transmitters, not PM. The PM transmitter has a natural 6-dB/octave rising characteristic, so this stage will make the FM signal equivalent to the PM signal. This characteristic serves to improve the signal-to-noise ratio of the system.

The audio signal from the microphone is amplified in a microphone preamplifier. The next stage is a peak limiter, also sometimes called a clipper. The limiting process tends to distort the signal, so a low-pass "splatter filter" is used following the peak limiter. Finally, the audio signal is amplified to the level needed to modulate the transmitter. The peak deviation is set by the peak audio amplitude, so a peak deviation control is essentially the gain or level control for the audio signal.

The difference between the FM and PM transmitter is found in the modulator section. Both can use a *reactance modulator*, but they differ in how it is applied. A reactance modulator produces an inductive reactance or capacitive reactance that varies with the audio signal. In the FM transmitter, the reactance modulator is applied directly to the frequency control circuits of the transmitter. In the PM transmitter, on the other hand, the reactance modulator is situated between the frequency source and the following stages. It is common practice to use PM when very

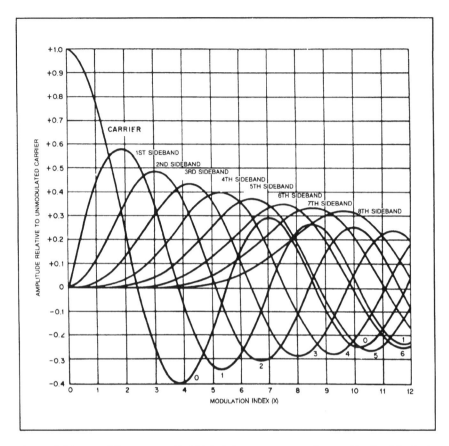

Figure 9.39 Bessell functions used in FM deviation measurement. (Source: *Handbook for Radio Amateurs.* Courtesy American Radio Relay League; Used by permission)

good frequency stability is needed because the reactance modulator in an FM transmitter can affect frequency stability and accuracy.

The following stages are frequency multipliers. These stages are used to generate frequencies that are integer multiples of the input frequency. There are two reasons for using this arrangement. First, because FM/PM transmitters are normally used in the VHF/UHF region, the master frequency source will be able to operate at a lower frequency where it is inherently easier to provide stability and accuracy. Also, the modulation process in FM and PM transmitters is linear only over a relatively small range of deviation. When wide deviation is attempted, the modulation distorts. But when a chain of frequency multipliers are used, the deviation is also multiplied.

For example, the transmitter in Figure 9.40 has a doubler and two triplers. The multiplication factor, therefore, is $2 \times 3 \times 3 = 18$. Suppose a transmitter operates on 161.575 MHz. The master oscillator frequency, therefore, must be 161.575 MHz/18 = 8.976389 MHz. If the deviation is supposed to be 5 kHz, the deviation at the master oscillator stage must be 5,000 Hz/18 = 277.78 Hz.

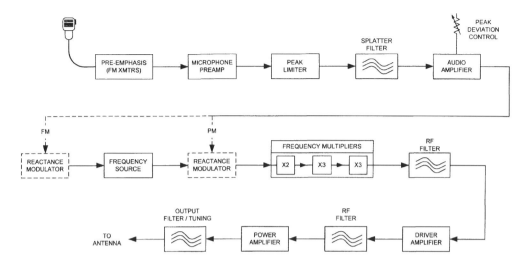

Figure 9.40 Frequency- or phase-modulated transmitter block diagram.

Frequency multiplication is a nonlinear process, so RF filtering must be provided following the mixer. It is a good practice to provide filtering in any transmitter in order to suppress any spurious emissions. The driver amplifier builds the signal to a level necessary to drive the final power amplifier. The output filter and tuning network serves to select the desired frequency while suppressing the spurs, and also to match the antenna impedance to the final amplifier.

Transmitter Test Set-up

Figure 9.41 shows a test set-up for FM transmitters. As with other types of transmitters, an adjustable power supply is used to power the transmitter. For mobile, aviation, and marine units, a variable DC power supply is used, while base station or fixed units can be powered from a variable AC transformer such as a *Variac®*.

The instrumentation set-up will vary somewhat with each situation, but the one shown in Figure 9.41 is representative. As with other types of transmitter, an RF wattmeter and dummy load are used. An isolated coupler or sampling tee is used to take a small portion of the signal to the test instruments. A *peak deviation meter* is a staple of FM transmitter servicing. Sometimes, an oscilloscope is also used (keep in mind that peak deviation is indicated by peak recovered audio). In many cases, either a spectrum analyzer or a specialized communications service monitor (or "test set") is also used.

Figure 9.42 shows a representative communications test set. This instrument is the Hewlett-Packard Model HP-8920A. It includes a spectrum analyzer with tracking generator, signaling encode/decode, and cellular radio test capability. It also include an AM/FM signal generator that outputs signals to 1,000 MHz, an AM/FM modulation analyzer, SSB demodulator, frequency counter, frequency error meter, RF power meter, distortion meter, SINAD receiver test meter, ac/dc

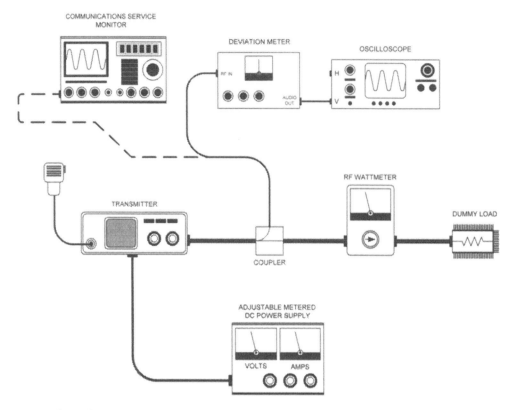

Figure 9.41 FM transmitter measurement test set-up.

voltmeter, digital oscilloscope, and other functions. It is microprocessor controlled and can output test results to a printer or a desktop computer.

RF POWER MEASUREMENT

The measurement of RF output power for FM/PM transmitters requires a standard averaging reading meter. Peak reading RF power meters are not well suited to this use. The usual procedure is to measure the RF power when no modulation is applied, as with AM transmitters. Experienced technicians will, however, measure the RF power both with and without modulation. The nature of FM/PM is such that the output power does not vary with modulation, so if a difference is noted it should be investigated.

Modulation Measurement

Modulation measurement with a peak deviation meter (as in Figure 9.41) is a straightforward process:

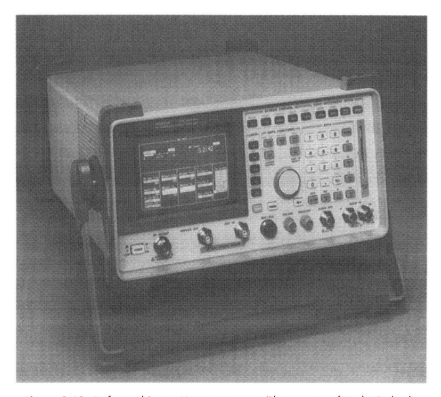

Figure 9.42 Professional Service Monitor instrument. (Photo courtesy of Hewlett-Packard)

1. Set-up the equipment and turn on the transmitter (make sure the power is as specified by the manufacturer of the transmitter).
2. Apply a 1,000-Hz sine wave to the microphone input of the transmitter. Start with a low-level signal and increase it until either the meter reading or the amplitude displayed on the oscilloscope cease increasing. In order to ensure proper clipping action, you might want to increase the audio amplitude a small amount more.
3. The peak deviation can be read from the deviation meter, or, if the oscilloscope vertical axis is properly calibrated in terms of deviation, from the 'scope screen.

The deviation meter is based on the fact that the discriminator circuit produces an output voltage that is proportional to the deviation (Figure 9.43A). The deviation meter is essentially a simple superheterodyne receiver with a discriminator for the demodulator. The output of the discriminator is fed to the deviation measurement circuits (Figure 9.43B). The absolute value (AV) circuit shown in Figure 9.43B is nothing more than a fancy fullwave rectifier. The output of the AV circuit is fed to a peak holding circuit to read peak deviation. In some instruments a *modulation density* function is provided that measures the average value of the full-wave rectified discriminator output.

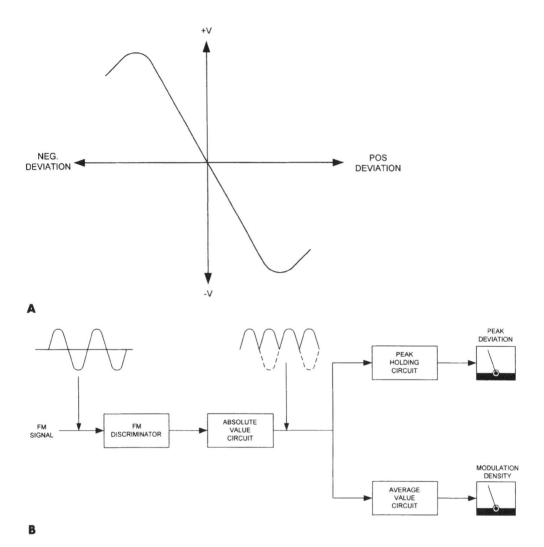

Figure 9.43 (A) FM discriminator output; (B) Simple deviation meter.

The deviation is both positive and negative because it swings above and below the carrier frequency. Ordinarily, the deviations above and below the carrier are equal, but sometimes a defect occurs to change this relationship. This produces *modulation dyssymmetry.* The percentage *of modulation dyssymmetry* is:

$$\% \ Dissymmetry = \frac{(Larger \ Deviation \ - \ Smaller \ Deviation) \times 100\%}{Larger \ Deviation} \quad [9.19]$$

Deviation by Bessel Zero Method

If you examine the FM transmitter spectrum (Figure 9.44), or the carrier amplitude-versus-modulation index sideband curves of Figure 9.39, you will note that the carrier amplitude goes through zero at certain critical points. The particular curves shown in Figure 9.39 are called Bessel functions. We can use these *Bessel zero points* either to make precision measurements of FM deviation or set the deviation to a specific level. The latter is often done on signal generators in calibration laboratories.

Because the modulation index is the ratio of peak deviation to modulating frequency, we can state that the peak deviation is the product of the modulation index and modulating frequency:

$$F_D = F_M \times \beta \qquad [9.20]$$

Table 9.1 shows a compilation of carrier nulls and standard deviations, along with the audio frequencies that will produce them. To use this data in the Bessel null method:

1. Pick the desired deviation from the horizontal column (e.g., 15 kHz).
2. Select a modulation index that corresponds to an audio frequency within the transmitter's audio passband. In this case, a modulation index of 8.65 occurs with a modulating frequency of 1.73 kHz.
3. Use an audio frequency counter of good accuracy to set the output frequency of a sine wave generator to exactly the required frequency (e.g., 1,730 Hz).
4. Increase the deviation of the carrier until the carrier has gone through one less than the number of nulls shown in the carrier null column of Table 9.1,

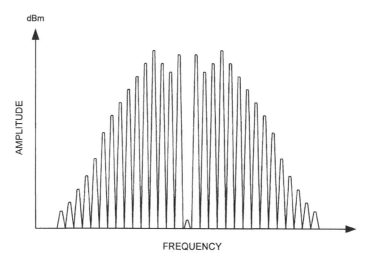

Figure 9.44 FM transmitter spectrum under sine wave modulation.

TABLE 9.1 Bessel null frequencies.

		Peak Deviation (kHz)									
Carrier Null	β	7.5	10	15	25	30	50	100	150	250	300
1	2.40	3.12	4.16	6.25	10.42	12.50	20.83	41.67	62.50	104.17	125.00
2	5.52	1.36	1.18	2.72	4.53	5.43	9.06	18.12	27.17	45.29	54.35
3	8.65	0.87	1.16	1.73	2.89	3.47	5.78	11.56	17.34	28.90	34.68
4	11.79	0.66	0.85	1.27	2.12	2.54	4.24	8.48	12.72	21.20	25.45
5	14.93	0.5	0.67	1.00	1.67	2.01	3.35	6.70	10.05	16.74	20.09
6	18.07	0.42	0.55	0.83	1.88	1.66	2.77	5.53	8.30	13.84	16.60

Audio frequencies in cells are in kilohertz (kHz)

and then stop on the next one. This level corresponds to the deviation shown in Table 9.1.

SPECIALIZED RF COMMUNICATIONS TEST SETS

We have already seen the communications test set, but that is only one form of instrument. Figures 9.45 and 9.46 show additional instruments. The Hewlett-Packard HP-8924 CDMA mobile station test set is shown in Figure 9.45. This instrument can be used to measure mobile and portable cellular and PCS telephone transmitters in the field, as well as in the plant. It includes a built-in spectrum analyzer, as well as the specialized features needed for CDMA (code domain multiple access)

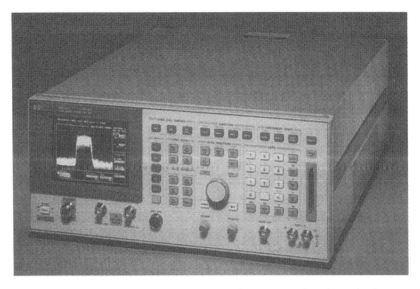

Figure 9.45 CDMA modulation analyzer. (Photo courtesy of Hewlett-Packard)

Figure 9.46 Modulation Domain analyzer. (Photo courtesy of Hewlett-Packard)

transmitters. It includes built-in signal generators and deviation meters for analog transmitters.

A Modulation Domain Meter is shown in Figure 9.46. This is a relatively new style of instrument. It has uses in a number of fields not related to RF communication, as well as applications in analyzing the modulation characteristics of radio transmitters.

CHAPTER TEN

Amplifier Measurements

Amplifiers are designed to amplify signals. Amplifiers fit into a number of different generic categories that define what they are and what they do. Certain concepts are common to all amplifiers, and still others are common to large classes of amplifiers using all forms of active devices (transistors, vacuum tubes, etc.) as the amplifying elements. The classifications of amplifiers as to the active element are well known. It is easy, for example, to see why some amplifiers are "vacuum tube amplifiers" and others are "transistor amplifiers." Oddly, in RF work vacuum tube amplifiers are not altogether obsolete, because some high-power transmitters still use tubes in the final stage. Transistor amplifiers are broken into *bipolar* (NPN and PNP) and *field effect transistor* (JFET and MOSFET) types.

We also have several other methods of classifying amplifier circuits, which are not dependent on the technology of the amplifying device. Several popular classification schemes are:

- Classification by common element
- Classification by active device conduction angle
- Classification by transfer function
- Classification by feedback versus nonfeedback
- Classification by frequency response

In this chapter we will consider some of the basics of amplifier circuits. Keep in mind that any of these could be constructed using transistors of either basic type, linear integrated circuits (ICs), or even vacuum tubes.

CLASSIFICATION BY COMMON ELEMENT

All of the circuits that you will study in this book have distinct input and output circuits that share at least one common point. Active devices used in amplifiers, such as transistors, are designed so that at least one element is common to both input and output circuits. In Chapter 11 you will discover that transistors have three elements: *collector, base,* and *emitter.* These form three different amplifiers: *common collector, common base,* and *common emitter.*

Classification by Conduction Angle

When one speaks of amplifier classes, it almost always means classification by conduction angle; that is, the portion of the input signal cycle over which the output current flows in the amplifying device. We recognize three traditional classes, which are labeled *Class-A, Class-B, Class-C,* and a combination class called *Class-AB.* There are also newer classes such as Class-E.

Class-A Amplifiers

In this class, the output current (in the *collector* element of bipolar transistors and the *drain* element in field effect transistors) flows over the entire 360° of the input cycle. This class is the least efficient class, but it is capable of low-distortion ("linear") operation using just one transistor.

Class-B Amplifiers

In this class, output current flows over 180° of the input cycle. This class is more efficient than Class-A amplifiers, but requires two or more transistors in order to make the amplifier linear.

Class-C Amplifiers

This class is the most efficient of the classes and has output current flowing over less than 180° of the input cycle. A typical conduction angle for a Class-C amplifier is 120°, although the range 90 to 150° can be found in practical examples. The Class-C amplifier is limited to RF power amplifier applications where linearity is not a requirement. Examples include the final RF power amplifiers in amplitude-modulated transmitters in which the modulation is applied to the final amplifier. Another example is FM transmitters where the information is carried in angular modulation form and amplitude is irrelevant.

The efficiency of an amplifier is dependent in part on the class of operation. Figure 10.1 shows a chart of amplifier efficiency versus class of operation and conduction angle. Efficiency is defined as the ratio of the DC input power consumed by the output circuit (collector path for bipolar devices) to the output power delivered to the load. Expressed as a percentage:

$$\eta = \frac{P_{OUT}}{P_{DC}} \times 100\% \qquad [10.1]$$

The Class-A amplifier does not operate with great efficiency. In fact, most Class-A amplifiers operate around 25% efficiency. These amplifiers, therefore, consume large amounts of DC power from the power supply in order to generate any appreciable amount of signal output power to the load. The rest of the power (about 75%) is dissipated as heat. As a result, Class-A amplifiers are used almost exclusively as voltage amplifiers, which deliver very little actual signal power, but do build up the voltage level considerably.

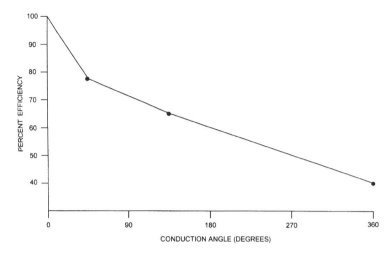

Figure 10.1 Amplifier efficiency as a function of conduction angle.

The lost power—the difference between the DC power consumed and the signal power delivered to the load—is given off as heat by the amplifier active device. Collector temperature is thus considerably higher in Class-A circuits than in the other circuits. Also, at zero-input signal when there is no input signal, the collector power dissipation is maximum. In the Class-B and Class-C designs, on the other hand, the collector power dissipation drops to near zero when no input signal is present.

The Class-B amplifier develops collector power only when the input signal is nonzero. This is because the collector current flows over only 180°, or one-half, of the input signal cycle. But it is impossible to make a linear amplifier using just one transistor in Class-B. It takes at least two transistors, driven 180° out of phase with each other so that they will operate on alternate halves of the input signal cycle, to make a Class-B amplifier.

The Class-C amplifier cannot be made linear, regardless of the number of transistors used. As a result, the Class-C amplifier cannot be used in applications where it is important to preserve the input waveform. There are no Class-C "hi-fi" audio amplifiers, for example. Similarly, there are no radio frequency (RF) amplifiers in Class-C if single sideband (SSB) or amplitude modulated (AM) signals are to be boosted. The Class-C amplifier is used for continuous wave (CW, or "morse code" telegraphy), frequency modulated (FM), or "high-level" AM signals. The latter are AM signals in which the Class-C amplifier itself is modulated, rather than a preceding stage.

Because Class-C amplifiers severely distort the output signal, they also generate a lot of harmonics of the input signal (i.e., 2F, 3F, 4F, where F is the input signal). In radio transmitters, the harmonics are a distinct disadvantage, so they are eliminated by inductor-capacitor "tank" circuits that remove the harmonics. In addition, the RF pulse signal produced by the short conduction angle is reconverted to a sine wave by the *flywheel effect* of the LC tank circuit.

Class-AB Amplifiers

The Class-AB amplifier circuits are an attempt (usually successful) to obtain some of the benefits of both Class-A and Class-B circuits. When two transistors are operated in the Class-AB "push-pull" configuration, the circuit proves more linear than Class-B and more efficient than Class-A. Some RF linear amplifiers are operated in Class AB.

Class-D and Class-E Amplifiers

These amplifiers are not strictly "conduction angle" classes. The Class-D amplifier is based on *pulse width modulation* (PWM) and the Class-E is a *switching mode* amplifier. Class-D amplifiers tend to be audio, but RF versions are operated into the low-medium wave range. Class-E amplifiers are operated well into the H.F. region.

CLASSIFICATION BY TRANSFER FUNCTION

The *transfer function* of any electronic network can be expressed as the *ratio of the output signal over the input signal.* For a *voltage amplifier,* for example, the transfer function is the ratio of the output signal voltage (V_O) over the input signal voltage (V_{in}). This ratio is often expressed as the *gain* of the circuit and is symbolized by A_V.

There are four different general subclasses of transfer function: *voltage, current, transconductance,* and *transresistance.* Most readers are already familiar with the first two classes (voltage and current).

The transfer function of voltage amplifiers is:

$$A_V = \frac{V_O}{V_{IN}}$$
[10.2]

Because the units in both the denominator and numerator are volts, and therefore cancel each other ("divide out"), the voltage gain units are dimensionless.

The transfer function of the current amplifier is:

$$A_I = \frac{I_O}{I_{IN}}$$
[10.3]

The *transconductance amplifier* and *transresistance amplifier* are nearly as well known to most readers, even though the former is used extensively in integrated circuits. The transresistance amplifier has a transfer function that relates an output voltage to an input current. The units of electrical resistance apply to the transfer equation of this form of amplifier, because in Ohm's law V/I is resistance (R); the unit of the transfer equation is the *ohm* (Ω). The transfer equation of a transresistance amplifier is:

$$R_M = \frac{V_O}{I_{IN}} \qquad [10.4]$$

The transconductance amplifier is only a little more familiar to most readers. Field effect transistors use transconductance as the gain expression.

Transconductance is the reciprocal of resistance (I/V). The older unit of transconductance was "ohm" reversed, or *mho.* Today, however, by international agreement the mho is renamed the *siemen* after an early pioneer in electrical science.

Any transconductance amplifier relates an output current to an input voltage:

$$G_M = \frac{I_O}{V_{IN}} \qquad [10.5]$$

You will see more about transconductance amplifiers when you study field effect transistors or operational transconductance amplifiers.

CLASSIFICATION BY FEEDBACK VERSUS NONFEEDBACK

Feedback is a sample of the output signal that is rerouted back to the input port of the amplifier. Feedback can make a mediocre amplifier act like it is a much higher-quality circuit. It is possible to cancel some distortion in the amplifier, and to make it a lot more stable, by using a little *negative feedback.*

The basic block diagram for a feedback amplifier is shown in Figure 10.2. The main amplifier will be a transistor or integrated circuit and will have an "open-loop" (i.e., without feedback) gain of A_{vol}. The output signal voltage, V_o, is sampled, and the sample is passed back to the amplifier input via a *feedback network.* The transfer function (or "gain") of the feedback network is designated by the Greek letter *beta* (β). In most cases, the gain of the feedback network will be negative, as the network is entirely made of passive components (resistors, capacitors, and inductors), and no amplification is provided (indeed, some loss is provided). Of course, it is possible to make a feedback amplifier in which there is a main forward amplifier and a feedback amplifier stage.

The feedback signal is summed with the source signal V_S to form an input signal $V_{in} = V_f - V_S$. The output signal is the product of this input signal and the voltage gain of the stage. In this section we are dealing with a voltage amplifier system, but we could just as easily call the circuit any of the other types of amplifiers by substituting the correct parameters.

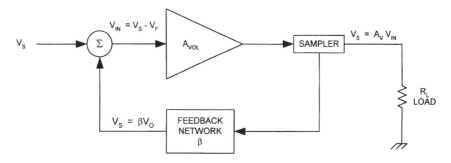

Figure 10.2 Block diagram of a feedback amplifier.

The gain of an amplifier with feedback is given by the expression:

$$A_V = \frac{A_{VOL}}{1 + \beta\,A_{VOL}} \qquad [10.6]$$

The use of feedback will improve the distortion situation of the amplifier. The *total harmonic distortion* (THD) will be reduced in the feedback amplifier, as in high-fidelity amplifiers used in audio. The degree of feedback is often expressed in decibels:

$$N_{dB} = 20\,LOG\left(\frac{A_V}{A_{VOL}}\right) = 20\,LOG\left(\frac{1}{1 + A_{VOL}\,\beta}\right) \qquad [10.7]$$

Figure 10.3 shows an example of a feedback network involving only resistors.

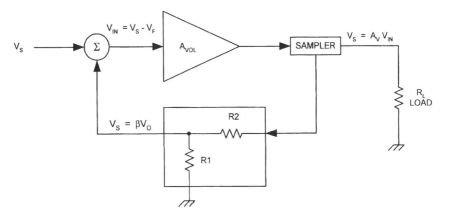

Figure 10.3 Feedback amplifier block diagram showing feedback network and voltages.

CLASSIFICATION BY FREQUENCY RESPONSE

Amplifiers are also classified as to approximate frequency response. This method is almost anecdotal, and there is considerable overlap of the various classes. One class is a *DC amplifier*. This type of amplifier will pass all AC frequencies within its range and DC levels. The fact that it is a DC amplifier does not mean that it won't pass AC signals, but that it *will* pass DC signals. An *audio amplifier* is one that will generally pass AC signals in the audio frequency range of roughly 30 to 20,000 Hz. A communications audio amplifier generally has a more limited audio response, such as 300 to 3,000 Hz. High-fidelity amplifiers, on the other hand, must have the full 20-kHz response, and many of them have a response of up to 50 kHz or 100 kHz. A *wideband amplifier* is exactly what its name implies; it presents a very wide bandwidth. You might find a wideband amplifier presenting a bandwidth of dozens or hundreds of kilohertz, or even megahertz. An amplifier that has a response up to 100 kHz or 1,000 kHz is surely a wideband amplifier. We would also call an amplifier with a 100-MHz bandwidth "wideband," but it is commonly the practice to refer to amplifiers with bandwidths greater than 1 MHz *video amplifiers*. This distinction is fading, however, as cable TV amplifiers are very wideband (2,000 MHz) but are called "wideband amplifiers" rather than "video amplifiers."

Radio frequency (RF) *amplifiers* may be tuned to some specific radio frequency (e.g., 155.123 MHz), or may be broadband over some specific band (154-158 MHz). These amplifiers are used in radio broadcasting and communications (receivers and transmitters) to select only the frequency, or band, of interest. The idea is to provide selection for the fundamental signal while discriminating against harmonics and spurious emissions far from the carrier (close-in spurs are not usually affected).

Amplifiers can also be characterized along a number of performance parameters:

- Frequency response (gain or phase versus frequency)
- Gain or phase versus power output
- −1 dB gain compression
- AM/PM conversion
- Input and output impedance matches
- VSWR
- Return loss
- Impedance values
- Maximum output power
- Total Harmonic distortion (THD)
- Intermodulation distortion
- Third-Order Intercept Point (TOIP)
- Noise figure
- Output-input isolation

The Small-Signal Amplifier

Small-signal amplifiers are typically used in receivers, preamplifiers, and some of the lower-level stages in transmitters. These amplifiers are often operated as Class-

A because of a need for linearity and the fact that heat dissipation from the inherent inefficiencies of Class-A are not a particular problem.

A typical small signal RF amplifier may be tuned to a single frequency, or broadbanded to cover an entire band of frequencies. Other amplifiers are extremely wideband. At one time only video amplifiers were so designed, but today monolithic microwave integrated circuits (MMIC) make it possible to produce wideband amplifiers with an inherent 50-ohms input/output impedance covering frequencies from DC to microwaves.

The RF Power Amplifier

RF power amplifiers are distinctly different from small-signal amplifiers. One obvious difference is the power levels involved. Where a small-signal amplifier, which may actually be a voltage amplifier, produces microwatts or a few milliwatts, the RF power amplifier may produce output power levels from 100 milliwatts up to several megawatts. A "license-free" transmitter, such as those used in portable telephones, may output only 100 mW, and portable cellular telephones about 600 mW. A few international shortwave radio stations, on the other hand, operate transmitters in the multimegawatt power class.

Another area of difference, and one that follows from the RF power level delivered, is the large amount of heat dissipated by the power amplifier. Efficiencies vary from 25% to less than 8%, so a tremendous amount of heat must be dissipated. This fact requires that RF power amplifiers use heat sinks and air blowers to remove the heat. Some high-power broadcast and other big transmitters use water cooling around the power amplifier's active device.

High-powered vacuum tubes used in RF amplifiers may use the grounded cathode-anode positive architecture, or they may ground the anode and supply a negative high voltage to the cathode. In the latter case, the anode structure is fitted with either a water jacket or air ducting system that is difficult to isolate if it is not grounded. It is not uncommon to find broadcast transmitter excess heat used to provide heat to the building in which it is housed in lieu of oil, gas, or a fuming manager.

Important RF Amplifier Parameters

Frequency Response

This parameter refers to the plot of gain versus frequency. The gain may be expressed in either linear units or decibels.

Gain/Phase vs. Power Output

The gain and phase shift (or phase error) varies with the output power level in RF power amplifiers.

−1 dB Gain Compression

An ideal textbook amplifier has a gain of P_{OUT}/P_{IN}, and both input and output can increase without limit. Real amplifiers begin to saturate at certain input power levels (Figure 10.4). At some point, just beyond the saturation knee, the gain will drop 1 dB. The input power level at which this change occurs is the −1 dB compression point.

AM/PM Conversion

AM/PM conversion is a measure of phase modulation (PM) components created by amplitude modulation (AM) variations of the signal. Sources of AM/PM conversion include both intentional and unintentional amplitude variations.

Impedance Match

The maximum power transfer, and freedom from reflections, occurs when the load and source impedances of a system are the same. The impedance of RF systems is nominally 50 ohms resistive (75 ohms in TV and video systems). This parameter indicates how well the input and output impedances of the amplifier match the system impedances.

VSWR or SWR

Whenever the source and load impedances are not matched, a standing wave ratio (SWR), also sometimes cited as voltage standing wave ratio (SWR), exists. The ratio of the two impedances (whichever produces a value greater than unity) is the VSWR. The VSWR of an amplifier port is VSWR that will exist at the port when it is connected to a 50-ohms source (input port) or load (output port). A VSWR of 1:1 is ideal.

Return Loss

Return loss is a measure of the reflected versus transmitted signal in a system, expressed in decibels (dB). As VSWR decreases towards 1:1 the return loss increases.

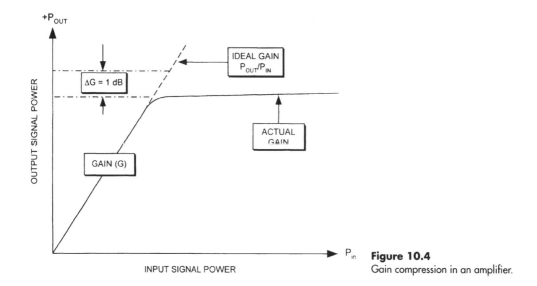

Figure 10.4
Gain compression in an amplifier.

Impedance Values

Impedance values are the actual value of input and output impedances, usually expressed in the form $Z = R + jX$.

Maximum Output Power

Maximum output power is used in RF power amplifiers to indicate the highest amount of power (in watts or its sub-units) that the amplifier can produce while still meeting specifications such as total harmonic distortion, intermodulation distortion, and so forth.

Total Harmonic Distortion (THD)

THD is a measure of linearity of an amplifier. A perfectly linear amplifier does not produce harmonics of the input frequency. Real amplifiers do, however, produce such harmonics (Figure 10.5).

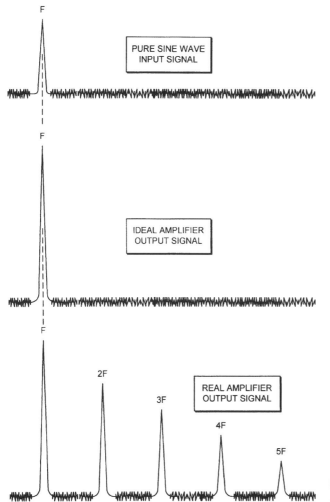

Figure 10.5
Ideal vs. practical output spectrum.

Intermodulation Distortion (IMD)

When two signals, F1 and F2, are simultaneously input to an amplifier that has any nonlinearity, then additional products will be generated. These are intermodulation distortion products. Figure 10.6 shows only the third-order products. They are usually regarded as the most important because they fall close to the original signals, so they will probably be within the amplifier passband. Other products may be outside the amplifier passband, so they are often of less (but still some) importance. The products are:

First-Order:	F1(original signals)
	F2
Second-Order:	F1 + F2
	F1 − F2
Third-Order:	2F1 + F2
	2F1 − F2
	2F2 + F1
	2F2 − F1
Fourth-Order:	2F1 + 2F2
	2F1 − 2F2
Fifth-Order:	3F1 + 2F2
	3F1 − 2F2
	3F2 + F1
	3F2 − F1

Et cetera . . .

Noise Figure

The noise figure, along with the noise temperature and noise factor, are measures of the amount of noise generated by the amplifier.

Third-Order Intercept Point (TOIP)

In an amplifier, the third-order products in the output increase 3 dB faster than the fundamental products (Figure 10.7). At some point, which may be above the −1 dB

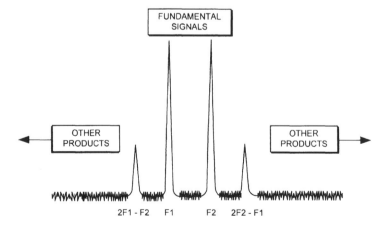

Figure 10.6 Intermodulation distortion (IMD) products.

compression point, the ideal fundamental response and the third-order response curve cross. This is the third-order intercept point. At input levels above the TOIP, both THD and IMD increases significantly.

Output-Input Isolation

Output-input isolation is a measure of the amount of signal applied to the output terminal of the amplifier that will appear in the input circuit. The applied signals might be reflected from the load (as when a power amplifier is connected to an antenna and transmission line), or they might be from following circuits (as in a receiver when the LO signal applied to the mixer may appear at the RF amplifier output).

Voltage Amplifier Gain Measurement

Figure 10.8 shows the test set-up for using a signal generator to make the voltage gain measurement. The signal generator must either have a well-calibrated output level control, or an RF voltmeter that is provided at the input. An RF voltmeter at the output terminal of the amplifier will read output voltage. Select a test frequency in mid-band and adjust the signal generator to that frequency. A test level is also

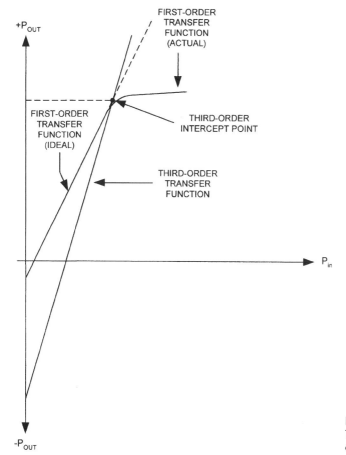

Figure 10.7
Third-order intercept point (TOIP or IP3).

needed, and is dependent on the amplifier properties. It is common practice to select an input level that is about 10 dB lower than the signal level that will produce −1 dB compression. The voltage gain of the amplifier is $A_V = V_{out}/V_{in}$. If the gain is to be expressed in decibels, then use the form $A_{VdB} = 20 \text{ LOG } A_V$.

Note, however, that the gain will typically vary over the passband, so select carefully. It is the practice of some people to make at least three measurements throughout the passband: low, high, and middle. These three are then either specified separately, or the results averaged.

1-dB Compression Point

This same set-up can be used in making 1-dB compression point measurements. The input level (either voltage or power) is increased in not more than 1-dB steps until the output level of the amplifier under test (AUT) does not increase with a 1-dB increase of input signal. Another way to measure it is to use a 10-dB input step, and to find the point at which the output only increases 9-dB.

Frequency Response Measurements

The frequency response measurement makes a plot of gain-versus-frequency. In this section we will take a look at several approaches.

Standard Signal Generator Method

Figure 10.8 also shows the signal generator method of making this measurement. The set-up is similar to the voltage gain system above, but can apply to either voltage or power response. In the case of power frequency response, the RF voltmeters are swapped for RF power meters. To make the measurement accurately, it is necessary to ensure that the signal level at the amplifier input is constant as the frequency changes. Many signal generators are equipped with *automatic level control* (ALC), which will do this task for you. But in other cases, the signal level will vary as the signal generator frequency is changed. In those cases the input voltmeter (or power meter) is used to monitor the levels. Two approaches can be taken. First, you can readjust the signal generator output to the test level (e.g., typically 10 dBm lower than the −1 dB compression point). Alternatively, you can measure and tabulate the input and output power levels and then perform the gain calculations.

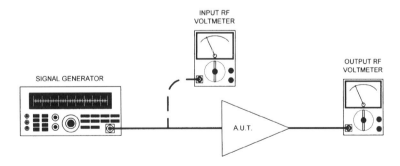

Figure 10.8 Simple gain measurement test set-up.

Sweeper Method

Sweep generators can also be used for the frequency response measurement. Figure 10.9 shows the test set-up. A sweep generator passes through a range of frequencies in one of two ways. First, it will sweep continuously through the range under the influence of a sawtooth waveform. Second, it might use a stepped-frequency approach in which the sweep generator output frequency changes a specified amount for each step. One way to accomplish this is to use a stepped sawtooth waveform such as is produced by a digital-to-analog converter that is driven by a binary counter.

Note in Figure 10.9 that the sweeper sawtooth (or stepped sawtooth) waveform is used to drive the horizontal input of the 'scope, while the output of the amplifier is used to drive the vertical input. If the sweeper is designed to maintain a constant output amplitude, then this trace will accurately reflect the frequency response of the amplifier being tested.

Spectrum Analyzer/Tracking Generator Method

A spectrum analyzer displays a plot of amplitude-versus-frequency. It works by sweeping the local oscillator (LO) of a superheterodyne receiver through a range of frequencies and then displaying the receiver output on an oscilloscope screen. A tracking generator is a special signal generator that is locked to the local oscillator and produces a signal equal to the center frequency of the analyzer's receiver passband. Some spectrum analyzers have internal tracking generators, while others (as shown in Figure 10.10) are external.

The general arrangement of Figure 10.10 is used in stimulus-response measurements, of which amplifier frequency response is an example. The tracking generator RF output is fed to the input of the amplifier under test (AUT), while the output of the AUT is fed to the receiver input on the spectrum analyzer. This produces a plot of the frequency response of the AUT.

Total Harmonic Distortion (THD) Measurement

Harmonic distortion exists when an amplifier is overdriven and begins to operate in the nonlinear region. Assuming that a pure sine wave signal, F1, is input to an

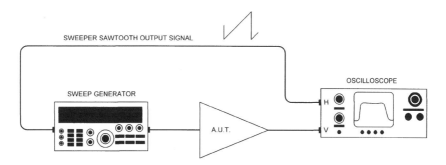

Figure 10.9   Sweeper used to measure amplifier frequency response.

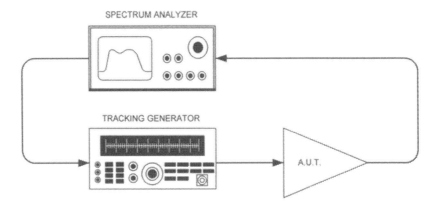

Figure 10.10 Spectrum analyzer and tracking generator used to measure amplifier frequency response.

AUT, the output spectrum of the ideal amplifier would be F1 only. But real amplifiers produce harmonics of F1, which fall at 2F1, 3F1, 4F1, 5F1, and so forth. The test set-up of Figure 10.11 will measure the Total Harmonic Distortion (THD) of the AUT. It consists of a signal generator with good spectral purity output, the AUT, and a notch filter set to eliminate frequency F1.

Switch S1 is initially in the position shown and measures V1. This voltage is the RMS sum of the fundamental and all harmonics. Next, switch S1 is switched to measure the output of the notch filter. This voltage, V2, is the RMS sum of all harmonics, less the fundamental (F1). The percentage THD is:

$$THD = \frac{V2 \times 100\%}{V1} \qquad [10.8]$$

Intermodulation Distortion Measurement

Intermodulation distortion (IMD) occurs when two or more signals mix in the amplifier input. If the amplifier is perfectly linear, then the output spectrum of the am-

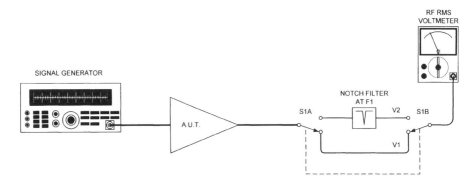

Figure 10.11 Basic concept for measuring Total Harmonic Distortion (THD).

plifier contains only the original input frequencies. But any nonlinearity at all produces the products discussed above. The usual method for characterizing the IMD performance of the AUT is to measure the second-order intercept point (IP2) and third-order intercept point (IP3).

Figure 10.12 shows the test set-up for making the IMD measurements. Two signal generators are provided. Each signal generator is equipped with a low-pass filter to ensure that the output spectrum of each generator contains no harmonic energy (which would produce IMD products of their own).

These signals, F1 and F2, are put together in a linear combiner network. It is important that the combiner have high port-to-port isolation to prevent the signal from one generator reaching the output port of the other. In some cases, variable or step attenuators are provided in these signal lines as well (which also helps isolation). The bandpass filter is used to ensure that no linearity problems prior to the amplifier input produce either harmonics or IMD products. Some combiners can produce nonlinearity, especially if ferrite or powdered iron core inductors are used.

The output of the AUT is displayed on a spectrum analyzer. Figure 10.13 shows the spectrum for IP2 measurement, while Figure 10.14 shows IP3 measurement. In both cases, equal amplitude (A) signals are applied at frequencies F1 and F2. In most amplifiers, a standard frequency difference between F1 and F2 is specified (e.g., 20 kHz, 100 kHz, 1 MHz, depending on amplifier types and frequency range).

IP2 Case

In the IP2 measurement we are looking at the second-order products, F1 ± F2. If the amplitude of the input signal is denoted A (dBm), and the gain is G (dB), then the output signal amplitude of the two equal amplitude input signals is $G + A$ dBm. The measurement is made by locating the second-order distortion products on the spectrum analyzer display and measuring the difference (in dB) between

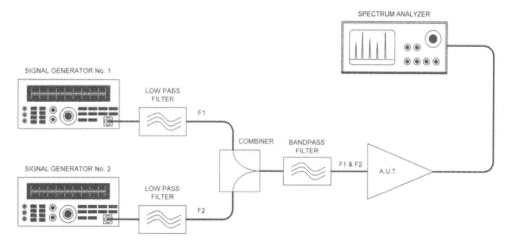

Figure 10.12 IMD measurement test set-up.

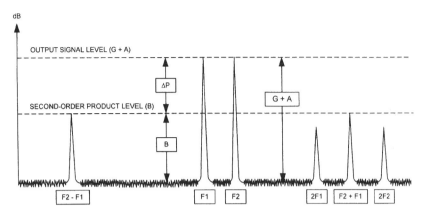

Figure 10.13 Spectrum for second-order IMD (IP2) measurement.

these products (ΔP) and the G + A output of the main signals (F1 and F2). The second-order intercept point is:

$$IP2 = A + \Delta P \qquad [10.9]$$

or,

$$IP2 = 2A + G - B \qquad [10.10]$$

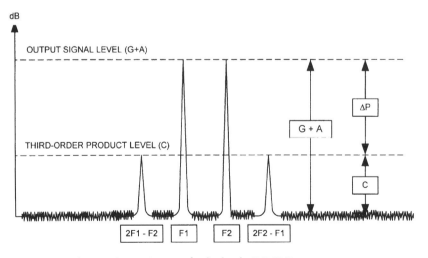

Figure 10.14 Spectrum for third-order IMD (IP3) measurement.

Where:

IP2 is the second-order intercept in dBm
A is the input signal level in dBm
G is the gain of the AUT in dB
B is the output level difference between the main signals and the second-order products

IP3 Case

The third-order intercept case (IP3) is measured in the same manner as IP2, but as the spectrum analyzer display in Figure 10.14 shows, we look at the third-order products. Of the four third-order products, the third-order difference products (2F1-F2 and 2F2-F1) are usually the most significant because they typically fall within the passband of the amplifier. The IP3 point is:

$$IP3 = A + \frac{\Delta P}{2} \qquad [10.11]$$

It is usually not necessary to measure higher-order intercept points.

RF Power Amplifier Tests

External RF power amplifiers are sometimes used to boost the output power of radio transmitters. Such amplifiers may have output power levels of a few watts or many kilowatts. Many of the tests used for transmitters also apply to these external power amplifiers, so you are advised to check Chapter 9, "Radio Transmitter Measurements." Figure 10.15 shows a typical set-up for measuring the performance of an external RF power amplifier. The exciter may be a signal generator or a low-power transmitter. In either case, the signal source must provide a signal on the correct frequency with a power level sufficient to drive the RF power amplifier.

Caution Note: Make sure that all components are capable of handling the power level, and that total isolating coupler and attenuator isolation reduces power to a level that the measurement instruments can accept!

There are two RF wattmeters in the system: one for the input and one for the output power. The dummy load is usually a resistive load capable of absorbing all of the power produced by the amplifier without overheating. In some cases, the dummy load will contain reactive components if the amplifier is intended to work into a reactive antenna, but that is rare.

The isolating coupler takes a sample of the output signal to be applied to either a spectrum analyzer or communications service monitor. In many cases, a separate variable or step attenuator is placed in the line between the coupler and the instruments.

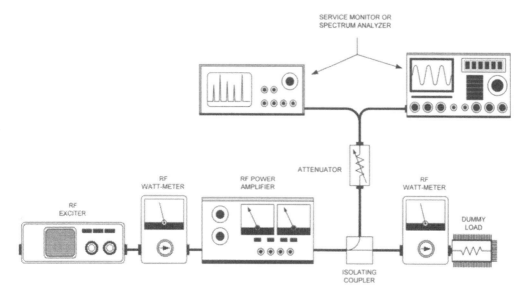

Figure 10.15 RF power amplifier measurement test set-up.

Gain Measurement

The gain of the amplifier is given by P_{Out}/P_{in}. To make this measurement it is necessary to use the RF wattmeters to measure the input power and output power. In some cases, the gain will be expressed in dBm, so the conversion must be made.

Output Power Level

The RF output power is often defined at a specified level of harmonic output (e.g., -40 dBc, or 40 dB below carrier). The input drive level is increased until the second harmonic reaches the specified level (assuming no input or output level restrictions are exceeded), and the output power is measured on the RF wattmeter.

In other cases, the output power level for a specified drive level is measured. In that case, the power level at the input RF wattmeter is set to the specified level, and the output power and second harmonic level are measured.

Note: Tests on RF amplifiers are often made using single-tone unmodulated CW signals. The actual performance with a modulated input signal may be different. For that reason, it is recommended that you also read Chapter 9, which deals with AM, SSB, and FM transmitters.

Simple Linearity Test

Figure 10.16 shows a method for performing linearity tests on an RF amplifier. Samples of the signal are taken from the input and output of the amplifier and applied to the vertical and horizontal inputs of an oscilloscope. If the RF amplifier operates at a higher frequency than the oscilloscope, then use envelope detectors for each sample. The resultant pattern on the oscilloscope screen is a variant of the Lissajous pattern.

Figure 10.17 shows representative traces. The trace in Figure 10.17A shows a perfectly linear amplifier, that is, the output/input plots as a straight line. In Figure 10.17B we see the case where either the amplifier is being overdriven or produces nonlinearity of its own. This trace indicates that the amplifier is going into saturation. In Figure 10.17C the trace indicates a possible improper bias condition in the amplifier.

Load-Pull Method

The load-pull approach to characterizing RF power amplifiers is shown in Figure 10.18. This procedure is done because large-signal RF power amplifiers often exhibit change as a function of output power level. The test set-up requires input and output tuners that are capable of providing different load conditions, then measuring the RF power and impedance at both the input and output of the amplifier. This procedure will help home in on input and output matching problems. The load-pull data are typically plotted on a Smith chart to show constant power levels over the entire range of impedances.

A variation on the theme is shown in Figure 10.19. This approach uses simultaneous drive of the input and output ports of the amplifier to simulate reflection from a load. Extreme care must be taken to prevent applying too high a level signal to the signal source.

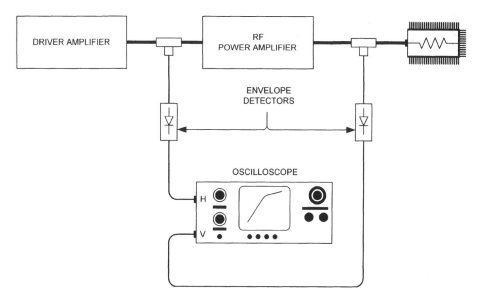

Figure 10.16 Test set-up for measuring amplifier linearity.

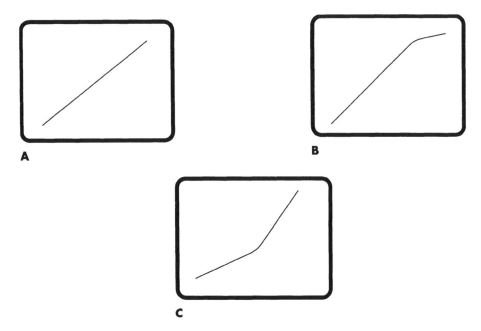

Figure 10.17 (A) Linearity good; (B) Amplifier is either being overdriven, or produces nonlinearity of its own (this trace indicates that the amplifier is going into saturation); (C) Possible improper bias condition in the amplifier.

Hot S_{22} Methods

The S-parameters are a powerful way to characterize amplifier (and other circuit) performance. In an RF power amplifier, the large-signal performance parameters vary as a function of the applied drive level. To measure all four S-parameters from the input side of the device we use a scheme such as Figure 10.20. Two loads ("A"

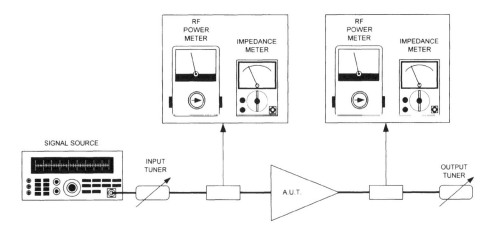

Figure 10.18 "Load-pull" approach to characterizing RF power amplifiers.

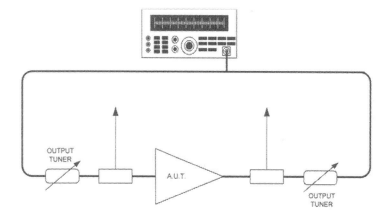

Figure 10.19 This approach to load-pull measurement uses simultaneous drive of the input and output ports of the amplifier to simulate reflection from a load.

and "B") are provided, and a network analyzer is used to measure incident and reflected power at the two ports. The four S-parameters are found from:

$$S_{11} = \frac{\left[\dfrac{b_0}{a_0} - \dfrac{b_0' \, a_3}{a_3' \, a_0} \right]}{\left[-\dfrac{a_3 \, a_0'}{a_0 \, a_3'} \right]}$$ [10.12]

$$S_{12} = \frac{\left[\dfrac{b_0'}{a_3'} - \dfrac{b_0 \, a_0'}{a_0 \, a_3'} \right]}{\left[-\dfrac{a_3 \, a_0'}{a_0 \, a_3'} \right]}$$ [10.13]

$$S_{21} = \frac{\left[\dfrac{b_3}{a_0} - \dfrac{b_3' \, a_3}{a_3' \, a_0} \right]}{\left[-\dfrac{a_3 \, a_0'}{a_0 \, a_3'} \right]}$$ [10.14]

$$S_{11} = \frac{\left[\dfrac{b_3'}{a_3'} - \dfrac{a_0' \, b_3}{a_3' \, a_0} \right]}{\left[-\dfrac{a_3 \, a_0'}{a_0 \, a_3'} \right]}$$ [10.15]

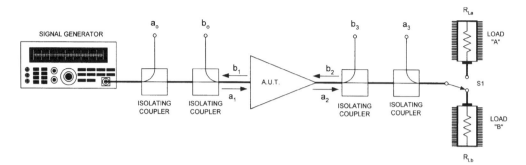

Figure 10.20 "Hot S_{22}" method of characterizing amplifier performance.

Where: the primed coefficients represent the measurements made with load "B" and the unprimed with load "A":

$Load$ "A":
$$b_0 = S_{11}a_0 + S_{21}a_3$$
$$b_3 = S_{21}a_0 + S_{22}a_3$$
$Load$ "B":
$$b_0' = S_{11}a_0' + S_{21}a_3'$$
$$b_3' = S_{21}a_0' + S_{22}a_3'$$

CHAPTER ELEVEN

Antenna Gain and Pattern Measurements

One way to characterize radio antennas is by their *radiation pattern.* When a signal propagates from an antenna it radiates in some directions more than others. A theoretical construct called an *isotropic source* is a spherical point source that radiates in all directions. An antenna shows an apparent increase in power in preferred directions because the power that is spread over the isotropic surface is concentrated in the preferred direction. This is the *gain* of the antenna. The gain of an antenna in any given direction is found essentially by comparing the signal strength in that direction to the signal strength of the same unit area on the surface of the radiating isotropic source. This is called *gain over isotropic* and is expressed in decibels (dBi).

When we make a map of the gains in all directions, we have the radiation pattern of the antenna. The overall radiation pattern is a solid three-dimensional geometric figure, so we often take horizontal ("azimuthal") and vertical ("elevation") slices out of the pattern and present them as representative of the whole.

Figure 11.1 shows the radiation pattern for a unidirectional Yagi-Uda array designed for 151 MHz. Ideally, the Yagi-Uda is unidirectional, but "real world" antennas are not so fortunate. In this case, there is a considerable back lobe and several side lobes in addition to the main lobe. Any lobe in an unwanted direction results in wasted power (lower signal strength on target), greater possibility of interfering with other stations, and (because of antenna reciprocity) more noise and interference on receive. Clearly, the antenna designer and the user need to know the nature of the azimuth and elevation patterns.

OUTDOOR TESTS

Figure 11.2 shows an outdoor method for measuring the antenna radiation pattern. It is fraught with difficulty, but is sometimes the only means available for a particular antenna or location. Although it is best to use an indoor antenna range (discussed below), those facilities are both expensive and frequency limited. All

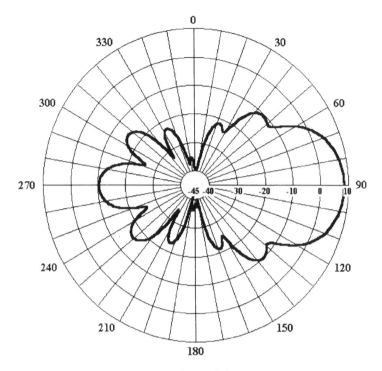

Figure 11.1 Typical azimuthal antenna pattern.

such facilities have a lower frequency cut-off limit, below which it will not work properly.

In Figure 11.2 the antenna under test (AUT) is mounted to a rotator assembly. A distant source, perhaps a distant on-the-air radio station, is used to provide a signal source. The signal source must be in the far-field, that is, where the power density falls off according to the inverse square law $(1/D^2)$. The pattern can be obtained by rotating the antenna through a 360° arc and measuring the signal strength at points along its arc of travel.

This process is easy, intuitive, and inherently appealing. But it is also problematic. For one thing, in the off-the-air variation on this theme, there is no guarantee that the output power of the distant radio station is constant during the time the antenna is rotating.

Some of the ambiguity can be taken out of the outdoor test approach by using a known reference signal, rather than an off-the-air signal. If a signal is generated, and controlled to maintain a constant output level, then part of the problem is solved. In general, the best approach to having the signal provide uniform illumination of the antenna is to radiate it from a location outside the near field of the antenna, from a transmitting loop antenna that is small compared with the wavelength of the signal being emitted. A loop that is $< 0.15\lambda$ has the attribute of exhibiting a constant current throughout the conductor. Larger loops, and other forms of antenna, have a varying current distribution along the length of the antenna radiator.

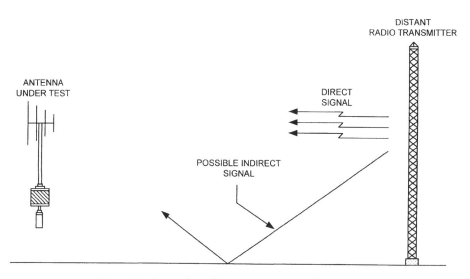

Figure 11.2 Simple outdoor antenna test range for patterns.

Also, unless it is a broadcasting station, it is likely to go on and off the air with use rather than emitting a constant signal. In the high-frequency shortwave bands this problem can be especially acute because of fading mechanisms afflicting even ground-wave signals. (With sky-wave sources—or ionospheric propagation, also known as "skip"—it is insanely difficult to account for!)

Another problem is unwanted signal paths. Normally, one hopes that only the direct space wave is received. That would make the problem easier. But in real environments there may be additional paths. In Figure 11.2 a ground reflection is shown. It is sometimes easy to overcome these by being aware of their source and adjusting the data accordingly. Unfortunately, nice, static ground reflections are not the only sort found. Especially at VHF/UHF, there may be transient reflections due to passing ground vehicles and aircraft. It is said that sometimes even good-sized birds can be a problem.

All in all, the outdoor method is perilous, at best. But before moving on to discussing the indoor-range approach, let's take a look at a quick and dirty approach to making gain measurements.

Figure 11.3 shows two antennas pointed at the same distant source. It is relatively easy to achieve a reasonable approximation of the antenna gain by comparing it to a reference antenna. For most purposes, either a half wavelength dipole or a quarter wavelength vertical antenna is preferred for the reference antenna.

The basic procedure is simple and straightforward. Let's assume that the AUT has a higher gain than the reference dipole. Set S1 to measure the signal strength coming from the reference antenna. Next, switch S1 to the AUT and again measure the signal strength.

If the receiver used has the ability to accurately measure the two signal levels, then a gain determination can be made. If the receiver is merely a communications receiver with an S-meter, then there may be problems if either: (a) the S-meter calibration scale factor is unknown, or (b) the scale factor changes over the range.

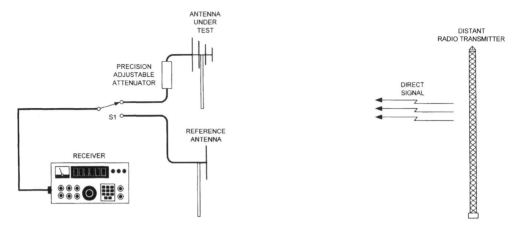

Figure 11.3 Outdoor test range for measuring gain.

(Some S-meters compress the upper end of the scale, so the difference between S8 and S9 is fewer decibels than the difference between, say, S1 and S2, despite the fact that both are "one S-unit"). In that case, a modified procedure is used:

1. Set S1 to measure the signal strength from the reference dipole.
2. Set S1 to measure the signal strength from the AUT (it should be higher).
3. Adjust the precision attenuator in the transmission line from the AUT until the S-meter reads exactly the same as it did for the reference dipole.

One instruction for this method took note of the fact that the signal levels might vary over time, so switching between antennas can introduce error. The problem is partially solved by switching the antenna at a slow enough rate to capture the data and taking a large number of samples (e.g., one per second over several minutes) and averaging the results.

INDOOR RANGE

If the funds and facilities are available, then an indoor antenna range (Figure 11.4) is preferred. The data collected are of far greater usefulness than the data collected out-of-doors, especially off-the-air. The indoor range is simply a more easily controlled, predictable environment.

The indoor range consists of a room lined with materials that absorb radio waves. Absorbers of various sizes, usually in the form of pyramid-shaped wedges, are placed over all surfaces of the chamber. The room is inside of a *Faraday cage*. This shielding provides protection against signals from outside interfering with the test, as well as reflections of the reference signal from structures outside of the chamber. A point source, such as a small loop at UHF and below, or a horn radiator at microwave frequencies, is positioned at one end of the room.

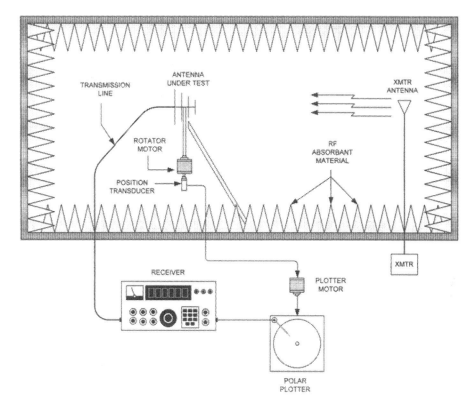

Figure 11.4 Indoor antenna anechoic test chamber.

The antenna is mounted on a rotatable pedestal. In order to prevent secondary reflections from the pedestal, it is often the case that blocks of absorbent material are used to block the view of the pedestal from the radiator. Both elevation and azimuth plots can be made by the simple expedient of mounting the antenna in the correct plane for each test.

The polar pattern of the antenna is created by using a servomotor or position transducer to rotate a chart recorder that measures the output strength of the receiver. The result is an antenna position-signal strength plot similar to the one shown in Figure 11.1.

Modern versions of the indoor range use a computer rather than the polar plotter shown in Figure 11.4. The output of the receiver is a voltage indicating signal level, so it can be A/D converted for input to the computer. The position transducer can also provide data to the computer. Once these data are in a plotting program, you can draw the polar plot.

One advantage of the computer approach is that a static profile of the chamber can be made by mounting a reference antenna in place of the AUT. The plot can then be made, and note taken of any reflections or other anomalies that exist in the chamber. These data can then be compared with the data for the AUT at each angle, with the AUT data adjusted to account for directional differences.

Antenna Modeling

Antenna chamber time can be expensive, so the trend today is to model the antenna performance on the computer. At one time, it was common to perform range tests and then tweak the antenna design to improve the performance (often with inconsistent results). Antenna modeling provides a lower-cost approach to tweaking performance, with range time reserved for final confirmation.

CHAPTER TWELVE

Antenna and Transmission Line Measurements

Antennas are used to convert the electrical signal produced by a transmitter into a transverse electromagnetic wave that can propagate through space. Although ideally the transmitter is located right at the feedpoint of the antenna, that is rarely the case. Most of the time a transmitter must be connected to its antenna through a transmission line (Figure 12.1). We need to be able to characterize the antenna and transmission line in order to be able to determine whether or not it is working correctly. Several factors interact to affect the antenna operation: transmitter, transmission line, and the impedance matching between the elements.

The transmitter will produce an RF output power level, P_O, on a specified frequency. This power is transferred to the antenna by way of the transmission line. All electrical source devices, including transmitters, exhibit an output impedance (Z_S in Figure 12.1). The maximum power transfer occurs only when the transmitter is connected to a load impedance (Z_L) equal to the source impedance. In a typical system, therefore, there are several opportunities for problems: the transmission line must be matched at one end to the antenna, and to the other end to the transmitter for maximum power transfer to occur.

Let's assume for the moment that the transmitter has a standard output impedance of 50 ohms and is connected to a transmission line, such as coaxial cable, with a characteristic impedance (Z_O) of 50 ohms. That leaves only the antenna to consider.

An antenna has a feedpoint impedance that depends on a number of factors. This impedance, Z_L, is of the form $Z_L = R \pm jX$. The jX component can be either inductive or capacitive reactance, although at resonance it will be zero. At frequencies below resonance, the antenna is too short to present a capacitive reactance, so $X = -jXC$. Similarly, if the antenna is operated at frequencies above resonance, it will exhibit inductive reactance, so $X = jXL$. For the time being we will consider the resonant case where X_L and X_C are equal, so balance out leaving $Z_L = R$.

If $Z_L = Z_O$, then all of the power presented to the feedpoint is absorbed by the antenna. In that case, no power is reflected back towards the transmitter. But if $Z_L \neq Z_O$ then not all of the applied power is absorbed by the antenna, some is

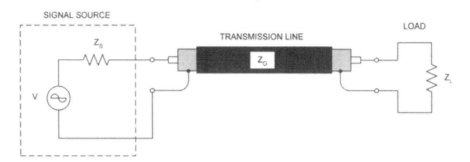

Figure 12.1 A transmitter must be connected to its antenna through a transmission line (all imped-ances matched).

reflected back towards the transmitter. At any point along the line the power ob-served will be a combination of the forward power PF and the reflected power P_R.

The ratio of the reflected power to the forward power is called the reflection coefficient (Γ). If you can measure the voltage along the line, then:

$$\Gamma = \frac{P_R}{P_F} \qquad [12.1]$$

We can also express the reflection coefficient in terms of the load and line impedances:

$$\Gamma = \frac{Z_L - Z_O}{Z_L + Z_O} \qquad [12.2]$$

The forward and reflected waves interfere and give rise to standing waves. A common, although often misleading, measure of the antenna system performance is the standing wave ratio (SWR) or, because it is often measured in terms of the voltages along the transmission line, voltage standing wave ratio (VSWR). For our purposes, VSWR and SWR are synonymous. A perfectly matched system has a VSWR of 1:1, while mismatched impedances are indicated by VSWR > 1:1. Some transmitters are designed to go into a protection mode when VSWR exceeds a cer-tain value. If the antenna presents a VSWR that is too high, output power is di-minished until it finally shuts off. As a result, with modern solid-state transmitters a good impedance match is more important than ever.

Figure 12.2 shows the VSWR situation a long the line at various lengths from the load end towards the transmitter. These plots show the measured voltage (P/Z) at points measured in terms of wavelength (λ). The matched case ($Z_L = Z_O$) is shown in Figure 12.2A. There is no reflected energy, so the voltage along the line (ignoring losses) is constant. Such a transmission line is said to be flat. Figures 12.2B and 12.2C show two extreme cases in which all of the forward power is reflected back towards the transmitter.

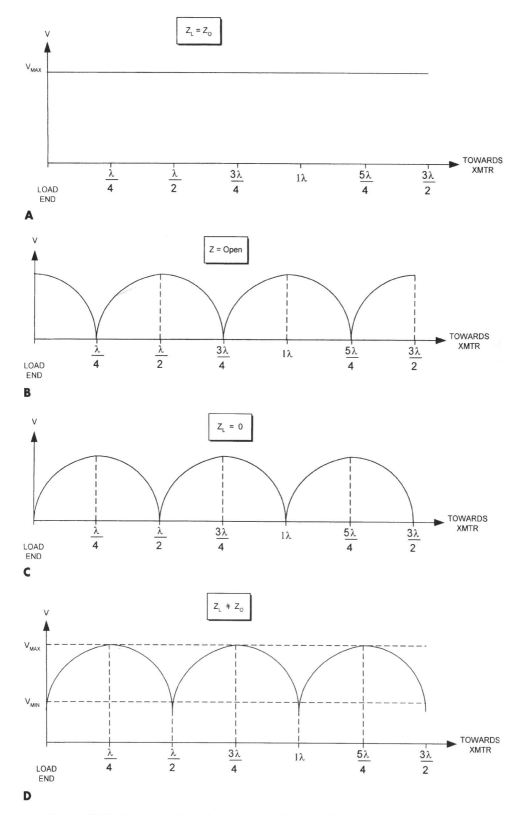

Figure 12.2 Transmission line voltage levels as a function of wavelength: (A) Impedances matched; (B) Load end open; (C) Load end shorted; (D) Impedances not matched, but neither open nor shorted.

In Figure 12.2B the line is open, so it is not connected to a load. All of the forward power arriving at the open load end of the line is reflected backwards toward the transmitter. The voltage is maximum at the load end and drops to zero quarter wavelength back down the line. The maxima (antinode) and minima (node) repeat each 180° (half wavelength) down the line, but 90° out of phase with each other.

The exact opposite situation is shown in Figure 12.2C. The load end of the transmission line is shorted ($Z_L = 0$). This means that a current antinode occurs at the load end, resulting in a voltage node. As in the previous case, the nodes and antinodes appear every half wavelength, but are displaced from each other by a quarter wavelength in each instance. Also, compare Figures 12.2B and 12.2C and note that there is a voltage node difference between the two cases. In Figure 12.2B the voltage nodes occur at integer multiples of quarter wavelength, while in Figure 12.2C they occur at the half wavelength multiples.

In most practical cases, short of a broken transmission line or other fault, the situation is not like either Figure 12.2B or 12.2C. A more common situation is as shown in Figure 12.2D. Here we have the case where $Z_L = Z_O$ and is neither zero nor infinite. There will be nodes and antinodes, and their relative positions depend on whether $Z_L > Z_O$ or $Z_L < Z_O$. The angle of the reflection also depends on these relationships. If $Z_L = Z_O$, then there is no reflection, so there can be no angular difference between the forward and reflected signals (the notion is meaningless). However, if $Z_L > Z_O$, then there is a 180° phase reversal between the forward and reflected signals. But if $Z_L < Z_O$, then there is no phase reversal.

Notice in Figure 12.2D that the node voltages do not go all the way to zero. The voltage along the line varies between V_{MAX} and V_{MIN}. From these values we can calculate the VSWR present in the system:

$$VSWR = \frac{V_{MAX}}{V_{MIN}} \qquad [12.3]$$

Or, knowing V_F and V_R:

$$VSWR = \frac{V_F + V_R}{V_F - V_R} \qquad [12.4]$$

We can also use the impedances to calculate VSWR:

$$\text{If } Z_L > Z_O: VSWR = Z_L/Z_O \qquad [12.5]$$

$$\text{If } Z_L < Z_O: VSWR = Z_O/Z_L \qquad [12.6]$$

The VSWR can also be determined by measuring the forward and reflected power levels (see Chapter 6, "Radio Frequency Power Measurements"):

$$VSWR = \frac{1 + \sqrt{P_R/P_F}}{1 - \sqrt{P_R/P_F}}$$ [12.7]

Or, if you substitute V2/Z for the powers, you will find that the reflection coefficient is the same form:

$$VSWR = \frac{1 + \Gamma}{1 - \Gamma}$$ [12.8]

To measure VSWR, then we can measure the impedances, the line voltage nodes and antinodes, the forward and reflected voltages, the power levels, or the reflection coefficient. All of these approaches have been used in VSWR measurement in one case or another.

When we make general measurements of antennas we are usually interested mainly in the VSWR and the resonant frequency. From these data we can infer quite a bit about the health of the antenna system. It does not, however, tell us anything about the radiation pattern of the antenna (see Chapter 11, "Antenna Gain and Pattern Measurements").

RF Power Meter Approach

An RF wattmeter can be used to measure the RF output of the transmitter, and if it is capable of distinguishing forward and reflected power it can also measure VSWR and infer resonant frequency. We know that the VSWR drops to minimum at resonance (ideally 1:1, but that is not a necessary condition to establish the resonant point). The resonant point is the point at which the reflected power drops to a minimum value.

Noise Bridge Approach

Figure 12.3A shows the basic noise bridge set-up, while Figure 12.3B shows the output signal from the bridge (note that it is a random gaussian signal). Figure 12.3C shows the simplified circuit of a typical noise-bridge instrument. The bridge consists of four arms. The inductive arms are in the form of a trifilar-wound transformer over a ferrite core, so signal applied to input winding is injected into the bridge circuit.

The measurement section consists of a series circuit of a 200-Ω potentiometer (R1) and a 250-pF variable capacitor (C1). The potentiometer sets the range (0 to 200 Ω) of the resistive component of measured impedance, while the capacitor sets the reactance component. Capacitor C2 in the "unknown" arm of the bridge is used to balance the measurement capacitor. With C2 in the circuit, the bridge is balanced

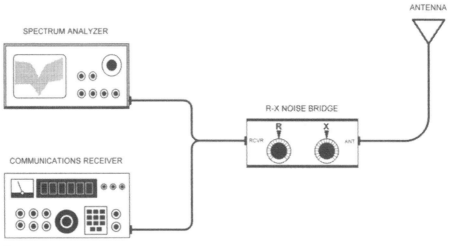

A

B

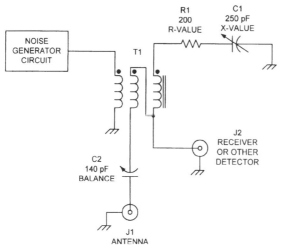

C

Figure 12.3
(A) Using the noise bridge;
(B) Noise signal; (C) Simple noise-
bridge circuit.

when C is in the center of its range. This arrangement accommodates both inductive and capacitive reactances, which appear on either side of the "zero" point (i.e., the midrange capacitance of C). When the bridge is in balance, the settings of R and C reveal the impedance across the ANTENNA terminal.

Adjusting Antennas with a Noise Bridge

Perhaps the most common use for the antenna noise bridge is finding the impedance and resonant points of an HF antenna. Connect the RECEIVER terminal of the bridge to the ANTENNA input of the HF receiver (or spectrum analyzer) through a short length of coaxial cable as shown in Figure 12.3A. The length should be as short as possible and the characteristic impedance should match that of the antenna feedline. Next, connect the coaxial feedline from the antenna to the ANTENNA terminals on the bridge. You are now ready to test the antenna.

Set the noise bridge resistance control to the antenna feedline impedance (usually 50 to 75 Ω for most common antennas). Set the reactance control to midrange (zero). Next, tune the receiver to the expected resonant frequency (F_{exp}) of the antenna. Turn the noise bridge on and look for a noise signal of about S9 (this will vary on different receivers), and if in the unlikely event that the antenna is resonant on the expected frequency, you will find yourself right in the middle of the null.

Adjust the resistance control (R1) on the bridge for a null, that is, minimum noise as indicated by the S-meter. Next, adjust the Reactance control (C1) for a null. Repeat the adjustments of the R1 and C1 controls for the deepest possible null, as indicated by either the lowest noise output on the S-meter or the deepest null on the spectrum analyzer (there is some interaction between the two controls).

A perfectly resonant antenna will have a reactance reading of zero ohms and a resistance of 50 to 75 Ω. Real antennas might have some reactance (the less the better) and a resistance that is somewhat different from 50 to 75 Ω. Impedance-matching methods can be used to transform the actual resistive component to the 50- or 75-Ω characteristic impedance of the transmission line. The results to expect:

1. If the resistance is close to zero, then suspect that there is a short circuit on the transmission line; and an open circuit, if the resistance is close to 200 Ω.
2. A reactance reading on the X_L side of zero indicates that the antenna is too long, while a reading on the X_C side of zero indicates an antenna that is too short.

An antenna that is too long or too short should be adjusted to the correct length. To determine the correct length, we must find the actual resonant frequency, F_R. To do this, reset the reactance control to zero, and then slowly tune the receiver in the proper direction (down-band for too long, and up-band for too short) until the null is found. On a high-Q antenna, the null is easy to miss if you tune too fast. Don't be surprised if that null is out of band by quite a bit. The percentage of change is given by dividing the expected resonant frequency (F_{exp}) by the actual resonant frequency (F_R) and multiplying by 100:

$$Change = \frac{F_{EXP} \times 100\%}{F_R} \qquad [12.9]$$

Connect the antenna, noise bridge, and receiver in the same manner as above. Set the receiver to the expected resonant frequency (i.e., approximately 468/F for half-wavelength types, and 234/F for quarter-wavelength types). Set the resistance control to 50 or 75 Ω, as appropriate for the normal antenna impedance and the transmission line impedance. Set the reactance control to zero. Turn the bridge on and listen for the noise signal.

Slowly rock the reactance control back and forth to find on which side of zero the null appears. Once the direction of the null is determined, set the reactance control to zero, and then tune the receiver toward the null direction (down-band if null is on X_L side, and up-band if it is on the X_C side of zero).

A less-than-ideal antenna will not have exactly 50 or 75 Ω impedance, so some adjustment of R1 and C1 to find the deepest null is in order. You will be surprised how far off some dipoles and other forms of antennas can be if they are not in "free space" (i.e., if they are close to the earth's surface).

Nonresonant Antenna Adjustment

We can operate antennas on frequencies other than their resonant frequency if we know the impedance:

$$X_C = \frac{55}{68 - C} - 2{,}340 \qquad\qquad [12.10]$$

or,

$$X_L = 2{,}340 - \frac{159{,}155}{68 + C} \qquad\qquad [12.11]$$

Plug the X value calculated from the equations above into $X_f = X/F$ where F is the desired frequency in MHz.

TRANSMISSION LINE MEASUREMENTS WITH NOISE BRIDGE

Some applications require antenna feedlines that are either a quarter wavelength or half wavelength at some specific frequency. In other cases, a piece of coaxial cable of specified length is required for other purposes: for instance, the dummy load used to service depth sounders is nothing but a long piece of shorted coax that returns the echo at a time interval that corresponds to a specific depth. We can use the bridge to find these lengths as follows:

1. Connect a short-circuit across the ANTENNA terminals and adjust R and X for the best null at the frequency of interest (note: both will be near zero).
2. Remove the short circuit.

3. Connect the length of transmission line to the UNKNOWN terminal—it should be longer than the expected length.
4. For quarter-wavelength lines, shorten the line until the null is very close to the desired frequency. For half-wavelength lines, do the same thing, except that the line must be shorted at the far end for each trial length.

The *velocity factor* of a transmission line (usually designated by the letter V in equations) is a decimal fraction that tells us how fast the radio wave propagates along the line relative to the speed of light in free space. For example, foam dielectric coaxial cable is said to have a velocity factor of $V = 0.80$. This number means that the signals in the line travel at a speed 0.80 (or 80%) of the speed of light.

Because all radio wavelength formulas are based on the velocity of light, you need the V value to calculate the physical length needed to equal any given electrical length. For example, a half-wavelength piece of coax has a physical length of $(492V)/F_{MHz}$ feet. Unfortunately, the real value of V is often a bit different from the published value. You can use the noise bridge to find the actual value of V for any sample of coaxial cable as follows:

1. Select a convenient length of the coax more than 12 feet in length, install a PL-259 RF connector (or other connector compatible with your instrument) on one end, and short-circuit the other end.
2. Accurately measure the physical length of the coax in feet; convert the "remainder" inches to a decimal fraction of one foot by dividing by 12 (e.g., $32'\ 8'' = 32.67'$ because $8''/12'' = 0.67$). Alternatively, cut off the cable to the nearest foot and reconnect the short circuit.
3. Set the bridge RESISTANCE and REACTANCE controls to zero.
4. Adjust the monitor receiver for deepest null. Use the null frequency to find the velocity factor $V = F_{MHz}L/492$, where V is the velocity factor (a decimal fraction); F_{MHz} is the frequency in MHz; and L is the cable length in feet.

Impedance Bridges

We can make antenna impedance measurements using a variant of the old-fashioned Wheatstone bridge. Figure 12.4A shows the basic form of the bridge in its most generalized form. The current flowing in the meter will be zero when $(Z1/Z2) = (Z3/Z4)$. If one arm of the bridge is the antenna impedance, then we can adjust the others to make the bridge null to make the measurement. A typical example is shown in Figure 12.4B. The antenna connected to J2 is one arm of the bridge, while R2 is a second. The value of R2 should be 50 or 75 Ω, depending upon the value of the expected antenna impedance. The choice of 68 Ω is a good compromise for meters to operate on both types of antennas. The other two arms of the bridge are the reactances of C1A and C1B, which is a single differential capacitor. Tune C1 until the meter is nulled, and then read the antenna from the dial. At least one instrument allowed the technician to plug in a resistor element equal to system impedance.

Calibrating the instrument is simple. A series of noninductive, carbon composition resistors having standard values from 10 to 1000 Ω are connected across J2. The meter is then nulled, and the value of the load resistor is inscribed on the dial at that point.

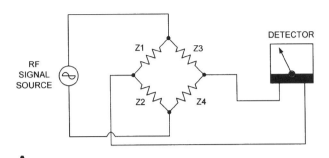

A

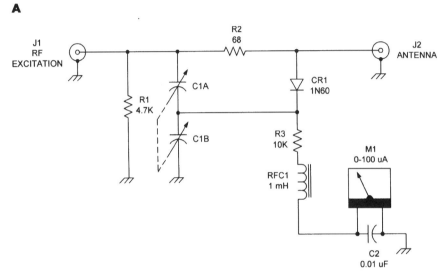

B

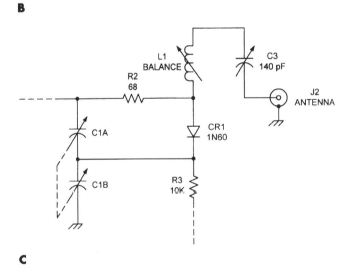

C

Figure 12.4
(A) Wheatstone bridge; (B) Practical
circuit based on the Wheatstone
bridge; (C) Modification for
reactance measurement.

The basic circuit of Figure 12.4B is useful only to measure the resistive component of impedance. We can modify the circuit as shown in Figure 12.4C to account for the reactive component.

VSWR BRIDGE METHODS

The VSWR bridge can be used to find the resonant point of antennas, as well as other parameters, in addition to the VSWR. Some instruments are metered, and are intended for the measurement of VSWR. Some also include RF power measurement, more or less. Figure 12.5 shows a type of bridge used for antenna and transmission line measurements. A typical in-line device (nonmetered) is shown in Figure 12.5A, and the stylized circuit including a directional coupler is shown in Figure 12.5B.

Impedance Bridge/VSWR Bridge

A traditional VSWR bridge is shown in Figure 12.5C. This instrument is used in-line to measure the VSWR of antenna systems. However, because of its design it cannot be left in-line for regular transmitter operations. The circuit of Figure 12.5C is basically a Wheatstone bridge consisting of R1 and the unknown impedance connected to the antenna port, J2, in two arms, and the combination R2, R3, and R4 in the other two arms. Diode detector D1 produces a DC output of the voltage a cross the bridge, so it will null when the bridge is in balance. By calibrating the dial of R3 in ohms, the resistive component of the unknown impedance can be determined by noting the resistance that balances the bridge.

Spectrum Analyzer Method

A spectrum analyzer can be paired with a regular signal generator to perform resonance tests of an antenna. Figure 12.6 shows the test set-up. The spectrum analyzer is connected to the RF OUT port of the VSWR bridge, while the antenna is connected to the DUT port. The idea is to look for the minimum height frequency spike on the spectrum analyzer. Set the spectrum analyzer to the correct frequency range so that it will accurately display the height of the signal at the antenna's resonant frequency. Tune the signal generator from a frequency away from resonance through the expected resonant point, while monitoring the amplitude on the spectrum analyzer screen. At some point, the spike will drop to a minimum value. The frequency at which this occurs is the resonant frequency of the antenna.

Tracking Generator Method

A tracking generator is a device that will follow the spectrum analyzer frequency as it sweeps. The tracking generator is heterodyned with the internal local oscillator (LO) of the spectrum analyzer to produce a signal of the same instantaneous frequency as is being displayed on the analyzer. As a result, the tracking generator can be paired with the spectrum analyzer to perform stimulus-response tests.

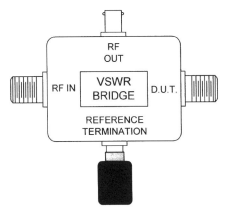

A

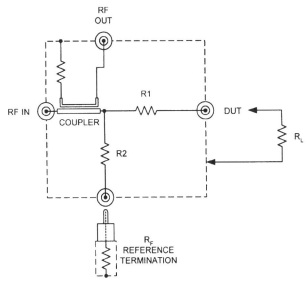

B

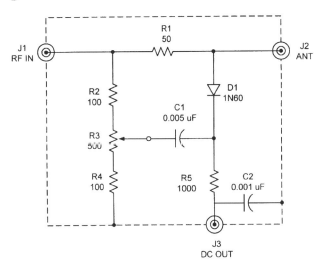

C

Figure 12.5
VSWR bridge: (A) Typical physical sensor form; (B) Notional circuit; (C) Standard bridge implementation.

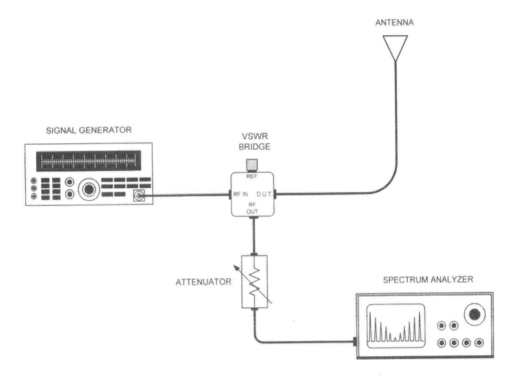

Figure 12.6 A signal generator and spectrum analyzer can be used to perform resonance tests of an antenna.

Figure 12.7 shows the connection of equipment needed to perform spectrum-analyzer/tracking-generator resonance tests of the antenna. The frequency at which the curve on the spectrum analyzer dips to a minimum is the resonant frequency of the antenna.

RF Voltmeter Method

Figure 12.8A shows the set-up for making resonance measurements with a signal generator and either an RF voltmeter or a DC voltmeter with envelope detector ("demodulator") probe. A sample circuit for a detector probe is shown in Figure 12.8B. The RF voltage at the RF OUT port of the VSWR bridge will drop to a minimum when the signal generator hits the resonant frequency of the antenna. Adjust the signal generator frequency slowly across the expected resonant frequency while monitoring the voltmeter. When the voltmeter dips to a minimum value, the resonance point is found.

Sweep Generator/Marker Method

A sweep generator will traverse through a range of frequencies under the influence of a sawtooth ramp. Figure 12.9A shows the set-up for using a sweep generator to test for antenna resonance. A *marker generator* is a crystal-controlled signal generator that outputs either a fundamental frequency and a rich collection of harmonics,

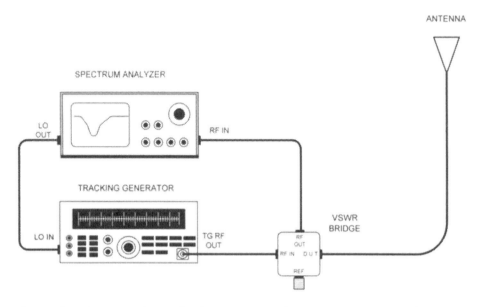

Figure 12.7 Spectrum analyzer and tracking generator used to measure antenna resonant frequency.

or a comb spectrum. The adder shown in Figure 12.9A nonlinearly combines the signals of the sweep generator (as passed through the VSWR bridge) and marker analyzer into one channel.

A typical adder is shown in Figure 12.9B. In some sweep generators, the marker generator and adder are built-in, so they are not needed externally. The composite marker and sweeper sample signal is demodulated by the envelope detector before being applied to the oscilloscope input.

Figure 12.9C shows a typical oscilloscope display for this test. The curve represents the amplitude of the sweeper signal, while the markers show up as vertical "pips" on the main curve. From these pips the frequencies are found. Sometimes, a variable marker signal generator is also used. The variable signal generator will be used to place a pip right at the minimum spot on the trough. The frequency of the signal generator at this point is the antenna's resonant frequency. The fixed markers can then be used to estimate the bandwidth of the antenna.

The "Low VSWR" Error

It is a common myth that a "low VSWR" is sufficient to know that the antenna and transmission line are in good order. But that's not true. Let's consider a situation where a high VSWR at the load end is reflected as a 2.05:1 at the input end of the line. This VSWR is well within specifications for many antenna systems. But is it accurate? Or is there an underlying problem? Consider an example. There is 250 feet (76.2-meters) of RG-8/U 52-ohms transmission line connecting the transmitter to the antenna. Suppose an antenna with a feedpoint impedance of

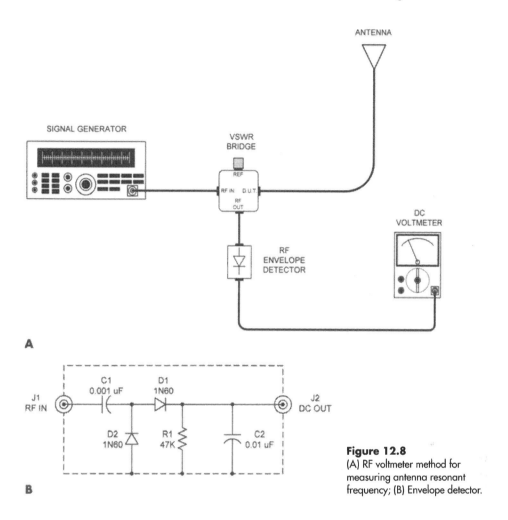

Figure 12.8
(A) RF voltmeter method for measuring antenna resonant frequency; (B) Envelope detector.

300-ohms resistive is connected to the other end of the RG-8/U coaxial cable. That would infer a VSWR of $300/52 = 5.77:1$, not the 2.05:1 that a technician would measure.

There are two basic forms of loss in coaxial cable: *copper loss* and *dielectric loss*. The copper losses result from the fact that the copper used to make the inner conductor and shield has resistance. The picture is further compounded by the fact that RF suffers skin effect, so the apparent RF resistance is higher for any given wire than the DC resistance. Further, the skin effect gets worse with frequency, so higher frequencies see more loss than lower frequencies.

The copper loss is due to the I^2R when current flows in the conductors. When there is a high VSWR at the load end, part of the power is reflected back down the line towards the transmitter. The effect of the reflected RF current is to increase the average current in the conductor. Thus, I^2R loss increases when VSWR increases.

This is also the case with dielectric loss. This loss occurs because voltage fields of the RF signal cause problems. The simplistic explanation is that the voltage fields

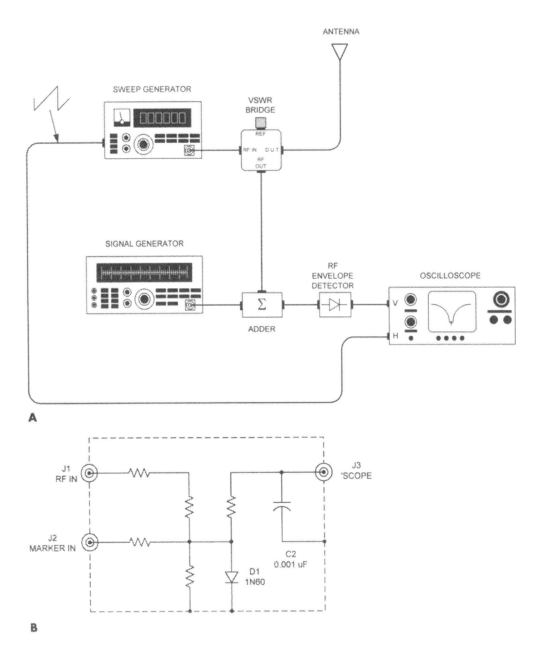

Figure 12.9 (A) Sweep generator used to measure resonant frequency; (B) Adder network; (C) Oscilloscope trace showing marker-generator "pips" for marking frequency.

tend to distort electron orbits, and when those orbits return to their normal state some energy is lost. These losses are related to V^2/R.

As with copper loss, the dielectric loss is frequency sensitive. The loss factor of coaxial cable increases with frequency. Let's look at two examples. Table 12.1 shows a popular, quality brand of two 52-ohms coaxial cables: RG-8/U and

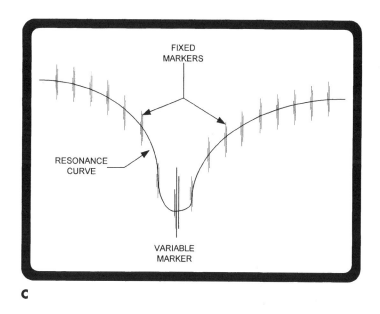

FIXED
MARKERS

RESONANCE
CURVE

VARIABLE
MARKER

c

RG-58/U. Note that the loss varies from 1.8 dB/100-feet (0.059 dB/meter) at 100-MHz to 7.10 dB/100-feet (0.233 dB/meter) at 1,000-MHz. The RG-8/U cable is larger than the RG-58/U and has less loss. Take a look at the same end points for the smaller cable: 4.9 dB/100-feet (0.161 dB/meter) at 100-MHz, and 21.50 dB/100-feet (0.705 dB/meter) at 1,000-MHz.

Clearly, the selection of the cable type is significant. At the FM BCB (100 MHz), the smaller cable would show 12.25 dB of loss. If a 100-watt signal is applied at the transmitter end, the ratio of loss is $10(12.25/10) = 16.79$, so only about 6 watts is available to the antenna. The problem is even worse if the antenna is used for 900-MHz cellular telephones. In that band, the RG-58/U cable loss is 20-dB/100-feet (0.656 dB/m), so the overall loss is a whopping 50 dB. The power ratio is $10^{(50/10)} = 10^5{:}1$, so only about 10 μW makes it to the antenna.

If RG-8/U cable is used instead of RG-58/U, then the losses would be 4.5 dB at 100 MHz and 16.75 dB at 900 MHz. At the cellular frequencies in the 900-MHz band the loss factor will still be high, about 47:1, so only 2 watts would make it through the loss. The rest is used to heat up the coaxial cable. Fortunately, certain specialty cables are available with losses around 2.5 dB/100-feet (0.082 dB/meter)

TABLE 12.1 Cable loss in decibels as a function of frequency.

	Frequency (MHz)									
Type	100	200	300	400	500	600	700	800	900	1,000
RG-8/U	1.80	2.70	3.45	4.20	4.73	5.27	5.80	6.25	6.70	7.10
RG-58/U	4.90	7.30	9.40	11.50	13.33	15.17	17.00	18.50	20.00	21.50

Loss in dB/100-feet

at 900 MHz. Such cable would produce about 6.25 dB of overall loss, or a ratio of 4.2:1. That cable would deliver nearly 24 watts of the original 100 watts.

The problem is also seen on receiver systems. Suppose that a 900-MHz receiver is at the end of a 250-foot transmission line. Further suppose that the signal is a respectable 1,000 μV, which in a 50-ohms load is −47 dBm. A loss of 6.25 dB would make the power level at the antenna terminals of the receiver −47 dBm − 6.25 dB = −53.25 dBm, or about 485 μV, which is still a reasonable signal. But if less coaxial cable is used (RG-58/U instead of the specialty grade), then the loss is 50 dB, and the signal at the receiver would be −47 dBm −50 dB = −97 dBm. This level is getting close to the sensitivity limits of some receivers, or about 3.2 μV.

The example given above was from telecommunications, but this effect can apply equally whenever RF is sent over coaxial cable. Cable TV, local area network, and other users of strictly landline RF also see the same loss effect. The correction is to connect a low-noise amplifier (LNA) at the head end of the transmission line. It boosts the signal before it suffers loss.

At first blush it might seem easier to put the amplifier at the receiver end. But that doesn't work out so well because of two factors. First, there is an inherent noise factor in any amplifier. If the signal is attenuated before it is applied to the amplifier, then the ratio of the signal to the internal noise of the amplifier is a lot lower than if the signal had been applied before attenuation. So, while the signal would still be at the same level regardless of where the amplifier is placed, the all-important signal-to-noise ratio is deteriorated if the amplifier is at the receiver end. The second reason is that any lossy device, including coaxial cable, produces a noise level of its own. The noise factor of a lossy device is:

$$F_N = 1 + \frac{(L - 1)\,T}{290} \qquad [12.12]$$

Where:

F_N is the noise factor of the coax
L is the loss express as a linear quantity
T is the physical temperature of the cable in Kelvins

The linear noise factor due to loss can be converted to noise figure, which can be added to the system noise decibel for decibel.

Now let's return to the problem of the "low" VSWR. The tables from the coaxial cable-maker tell us that the loss at 40 MHz is 1.2 dB/100-feet, so the overall loss is 3 dB (halving the power). This loss is called the *matched line loss* (L_M). But, we also have to consider the *Total Line Loss* (TLL), which is:

$$T.L.L. = 10\,LOG\left[\frac{B^2 - C^2}{B(1 - C^2)}\right] \qquad [12.13]$$

Where:

$B = $ Antilog L_M
$C = (SWRLOAD - 1)/(SWRLOAD + 1)$
SWRLOAD is the VSWR at the load end of the line

We know that the VSWR measured at the load end might be considerably higher than that measured at the transmitter end of the line. Given that $L_M = 3$ dB, $B = LOG^{-1}$ (3) $= 1.995$. If $VSWR_{LOAD} = 5.77$ (as is the case), then $C = (5.77 - 1)/(5.77 + 1) = 4.77/6.77 = 0.705$. Thus, the T.L.L. is

$$T.L.L. = 10 \, LOG \left[\frac{(1.995)^2 - (0.705)^2}{(1.995)(1 - 0.705^2)} \right] = 5.4 \, dB \qquad [12.14]$$

The VSWR at the input end of the line, down the hill by the transmitter, is then:

$$VSWR_{INPUT} = \frac{B + C}{B - C} = \frac{1.995 + 0.705}{1.995 - 0.705} = \frac{2.7}{1.29} = 2.05{:}1 \qquad [12.15]$$

It's clear from the above that a VSWR close to 6:1 could reflect to the transmitter end at 2.05:1. Unfortunately, the value 2.05:1 is well within specifications for many transmitter-antenna systems and therefore might not be seen as a problem by many technicians.

Return Loss Methods

Return loss is the comparison of the forward and reflected signals, as expressed in decibels (dB). It compares an unknown impedance with a known impedance, and then reports the degree of mismatch or match in both phase and magnitude. When the match is perfect, then the return loss is infinite, and when all of the forward power is reflected (as in a shorted or open load) the return loss is minimized ($P_F = P_R$). In addition, the same method is used for characterizing filters and other load-sensitive circuits.

Power Meter Method

The definition of return loss is given by Equation 12.16:

$$Return \, Loss = 10 \, LOG \left[\frac{P_F}{P_R} \right] dB \qquad [12.16]$$

This equation suggests a simple method for measuring return loss: compare the forward and reflected power using a directional RF wattmeter (see Chapter 6, "Radio Frequency Power Measurements"). The actual return loss of the load is the indicated return loss less twice the known transmission line loss (when return loss is measured at the transmitter end of the line).

Return loss can also be used to determine reflection coefficient and VSWR. We can write the return loss expression as:

$$Return \ Loss = 20 \ LOG \ |\Gamma| \qquad\qquad [12.17]$$

To determine reflection coefficient we need only solve Equation 12.17 for Γ:

$$|\Gamma| = 10^{(RL/20)} \qquad\qquad [12.18]$$

Where:

Γ is the voltage reflection coefficient
RL is the return loss in decibels (dB)

To determine VSWR from knowledge of the return loss:

$$VSWR = \frac{1 + 10^{(RL/20)}}{1 - 10^{(RL/20)}} \qquad\qquad [12.19]$$

Return Loss Bridge Method

A *return loss bridge* can be made using a hybrid combiner circuit (Figure 12.10A). All three resistors in the bridge are set to the system impedance, which is usually 50 ohms. Figure 12.10B shows the connection scheme for using the bridge to measure return loss. Initially, the unknown impedance is not connected to the UNKNOWN IMPEDANCE port (J2) of the return loss bridge. An RF signal source is tuned to the frequency at which the measurement is made.

The signal source might be a signal generator, a special high-power source, or a transmitter. The idea is to deliver enough RF power to the RF power so that small levels of reflected signal power can be measured on the selected meter. A step attenuator is in the line between the RF POWER METER port (J3) of the bridge and the RF wattmeter. Set the power level from the source so that a relatively large amount of attenuation is required to bring the wattmeter to some value in the upper half of its scale.

With the unknown impedance disconnected, measure the RF power appearing at the RF POWER METER port (J3). Once this value is recorded connect the unknown load impedance (which might be an antenna) and again measure the RF power. When measuring this reflected power it may prove necessary to adjust the

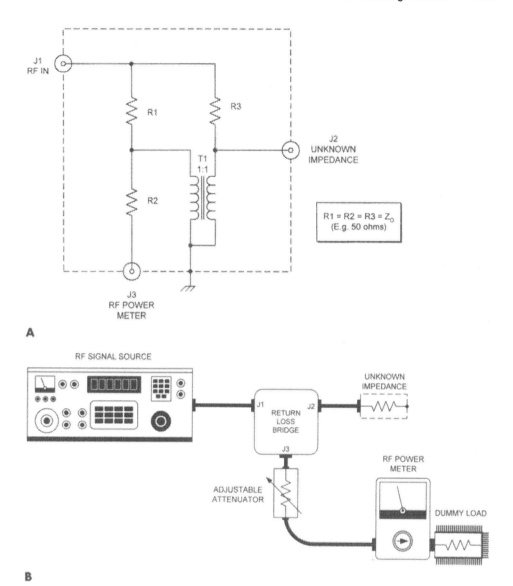

Figure 12.10 (A) Return loss bridge circuit; (B) Using the return loss bridge.

step attenuator to get a good reading. The value of the attenuation must be ac-counted for when making this measurement.

An equal deflection method could also be used in this measurement. Set the attenuator to a high value that produces a reading at a recognizable point on the meter (e.g., a calibration point, full-scale, or some other repeatable point). Note the amount of attenuation required. Connect the unknown impedance to J2 on the bridge, and the power indication on the meter will drop. Adjust the attenuator until the meter again reads the same value as before. The difference in the settings of the attenuator required to make the two readings is the return loss.

TRANSMISSION LINE LOSS

Transmission line losses occur because of resistive losses in the conductor and dielectric losses in the inner insulator. These losses are usually measured in terms of decibels per 100 feet (dB/100 ft), decibels per 100 meters (dB/100 m), or decibels per meter (dB/m). The latter is especially likely to be used in the high microwave bands.

Directional RF Wattmeter Method

The directional RF wattmeter can be used to measure transmission line loss by connecting a transmitter (or other RF power source) to a very long transmission line (several wavelengths). The far end of the transmission line is left unterminated (do *not* connect a load). Connect the directional RF wattmeter right at the transmitter end of the line. Measure the forward and reflected power and use them to calculate loss:

$$Loss\ (dB) = \frac{10\ LOG\ [P_F/P_R]}{2} \qquad [12.20]$$

To calculate the *loss factor*, that is, the loss per unit length (e.g., loss/100 feet), multiply the total loss by 100 and divide by the actual line length in feet:

$$Loss\ Factor = \frac{Total\ Line\ Loss \times 100}{Actual\ Line\ Length\ in\ Feet} \qquad [12.21]$$

MEASURING TRANSMISSION LINE VELOCITY FACTOR

The *velocity factor* of a transmission line is a measure of the velocity of propagation of signals in the line compared to the speed of light ($V = v/c$). The velocity factor is always a decimal fraction because $v < c$.

Figure 12.11 shows the test set-up for using a sweep generator to measure the velocity factor of a section of transmission line of known length. It is important that the transmission line be left unterminated. The pattern on the oscilloscope screen (Figure 12.12) is amplitude-versus-frequency by virtue of the fact that the sawtooth waveform that sweeps the signal generator also sweeps the oscilloscope X-axis. If we know the length of the line L, then we can measure the velocity factor from:

$$V = \frac{\Delta F \times L}{K}$$

[12.22]

Where:

V is the velocity factor $(0 - 1)$

L is the length

K is 492 if L is in feet and 150 if L is in meters

ΔF is the difference in frequency between either nodes or antinodes

A regular signal generator with accurately calibrated output frequency can be used to measure the ΔF factor. It is used as a marker and is added to the sweeper signal in an adder network. Position the pip from the marker over either a node or antinode and measure its frequency (F1). Then go to the next node or antinode and measure its frequency (F2). The term $\Delta F = |F2-F1|$.

A practical tip in making this measurement is to select the node rather than the antinode, because the nodes are sharper than the antinodes. As a result, it is easier to see the exact point at which they cross the X-axis.

It is also a good idea to measure ΔF twice and then average the results. For example, we might measure the frequency of the first node on the left side of the

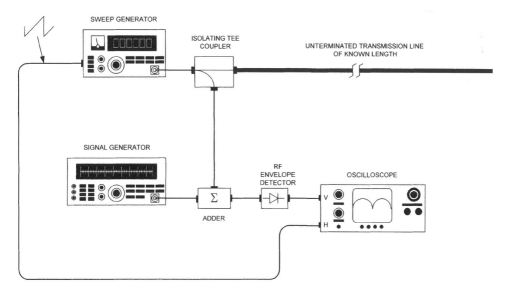

Figure 12.11 Test set-up for using a sweep generator to measure the velocity factor of a section of transmission line of known length.

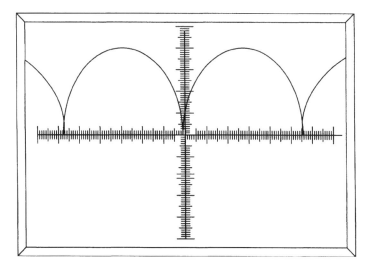

Figure 12.12 Amplitude-vs.-frequency display is because the sawtooth waveform that sweeps the signal generator also sweeps the oscilloscope X-axis.

screen (F1), the frequency of the second node (F2), and the frequency of the third node (F3). The average value is usually more accurate due to measurement errors:

$$\overline{\Delta F} = \frac{|F2 - F1| + |F3 - F2|}{2} \qquad [12.23]$$

FINDING A TRANSMISSION LINE FAULT

It is often necessary for a technician to find the distance to a fault on a transmission line that causes either a short circuit or an open circuit. The velocity factor method can be used for this purpose if Equation 12.22 is solved for L rather than V, and V is known from the manufacturer's data. Alternatively, you can also use *time-domain reflectometry* (see Chapter 14, "Time-Domain Reflectometry").

Tuning Want Test

Sometimes, the information that we need to adjust an antenna is whether it is too long or too short. Either case will produce a high VSWR. If the transmitter (or a signal generator) can be varied above and below the desired resonant frequency of the antenna, then we can observe whether the minimum VSWR is below or above the desired point. If it is below the desired frequency, then the antenna is too long, and if it is above, the antenna is too short. But what of the case where the transmitter is on a fixed frequency? In that case an old radio technician's trick can be used for VHF mobile units: the tuning wand (Figure 12.13).

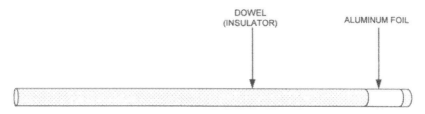

Figure 12.13 Test wand for testing VHF/UHF mobile antennas.

A tuning wand consists of an insulating dowel (wood, Lucite, Plexiglas, etc.) around one end of which a 4-inch length of aluminum foil is wrapped (for low-band VHF a longer foil section may be needed, and for higher UHF a shorter section may be needed).

To make a tuning-wand measurement, first measure the selected power (or VSWR) with the rod away from the antenna. Next, place the aluminum foil close to the antenna (a few millimeters), without touching it, at a point about midway up the antenna. Slowly raise the wand tip towards the antenna tip, while monitoring the VSWR or reflected power:

1. If the reflected power (or VSWR) goes up, then the antenna is too long.
2. If the reflected power (or VSWR) goes down, then the antenna is too short.

Field Strength Measurements

The field strength of the signal emitted by a radio antenna is usually measured in terms of volts per meter (V/m), or the subunits millivolts per meter (mV/m) or microvolts per meter (μV/m). Of these measures, the latter (μV/m) is the most commonly used in radio work.

The test set-up shown in Figure 12.14 is based on finding the antenna signal power in dBm and then converting it to μV/m. Two indicator instruments are shown in Figure 12.14, although in actuality only one will be needed. One indicator is a radio receiver with an S-meter to indicate signal strength. The exact calibration is not important, but what is important is the ability to repeat settings of the meter. The other instrument is a spectrum analyzer calibrated in decibels.

A test antenna is used to pick up the signal. It is important that the gain (G) of this antenna be known so that we can calculate an antenna factor, K:

$$K = (20 \, LOG \, F) - G - 29.8 \qquad [12.24]$$

Where:

K is the antenna factor in decibels (dB)
F is the frequency in megahertz (MHz)
G is the antenna gain in decibels (dB)

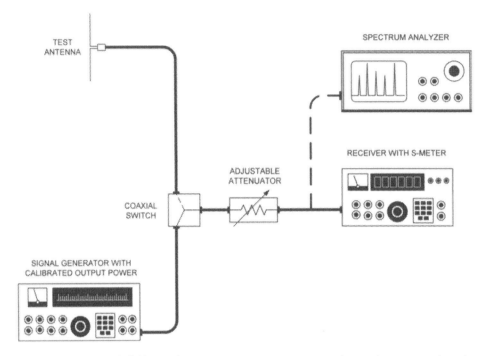

Figure 12.14 Simple field strength measurements use a test antenna and a signal generator with a calibrated output level.

For example, suppose we have an antenna with a gain of 3 dB used to measure the field strength of a signal on 155 MHz. The antenna correction factor is

K = (20 LOG 155) − 3 dB − 29.8 dB
K = 43.81 − 3 − 29.8 = 11.01 dB

To find the power, center the antenna on the radiated signal (plane of antenna orthogonal to the direction of propagation), and note the S-meter reading (use a small amount of attenuation in the adjustable attenuator to set the S-meter right on a tick mark). Next, turn the coaxial switch to the signal generator and tune the generator to the same frequency as the receiver. Adjust the output level of the signal generator until the same deflection of the S-meter is noted. The calibrated output power of the signal generator is equal to the signal power level at the receiver input.

The next chore is to convert the signal level in dBm to dBμV by the simple expedient of adding 107 dB to the reading (assuming 50 ohms impedance).

$$S_{dB\mu V} = P_{dBm} + 107 \text{ dB} \qquad [12.25]$$

Suppose, for example, the signal level was found to be −30 dBm. The signal level in $S_{dB\mu V}$ is −30 dBm + 107 dB = 77 dBμV.

Finding the signal level in dBμV/m, add the antenna correction factor K to the signal level in dBμV:

$$dB\mu V/m = S_{dB\mu V} + K \qquad [12.26]$$

Or, in the case of the 155 MHz signal that produced a −30 dBm signal: 77 dBμV + 11 dB = 88 dBμV/m. Finally, find the field strength in μV/m:

$$\mu V/m = LOG^{-1} \left[\frac{dB\mu V/m}{20} \right] \qquad [12.27]$$

In the case of our example, where the field is 88 dBμV/m, it is LOG^{-1} (88/20) = LOG^{-1}(4.4) = 25,119 μV/m.

Hand-Held Antenna Analyzers

Figures 12.15 and 12.16 show two different hand-held antenna analyzers. The unit in Figure 12.15, MFJ Enterprises Model MFJ-259B, was intended for the amateur radio market, but it is also useful for any frequency up to 170 MHz. It has a built-in signal source and frequency counter to ensure that the correct frequency is used. It has side-by-side VSWR and complex impedance analog meters. It will measure complex impedance $Z = R \pm jX$, or as magnitude and phase angle. It measures VSWR, return loss, and reflection coefficient at any frequency within its range. In addition, it can measure capacitance and inductance values.

The Bird Electronics Model AT-800 is shown in Figure 12.16. This particular model measures antenna performance over the 806 to 960 MHz range (other models test other bands). It will sweep the frequency range set and graph the VSWR on the LCD panel.

Figure 12.15
Low-cost VSWR analyzer
instrument. (Photo courtesy of MFJ
Enterprises, Inc.)

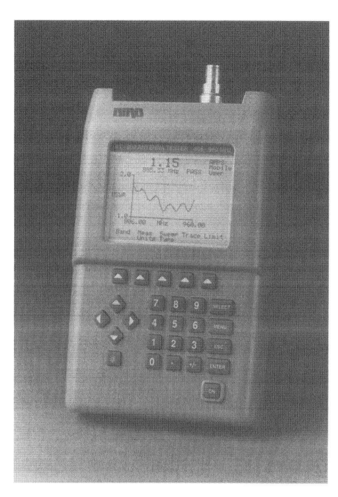

Figure 12.16
Graphing VSWR antenna analyzer.
(Photo courtesy of Bird Electronics
Corporation)

CHAPTER THIRTEEN

Measuring Inductors and Capacitors at RF Frequencies

The measurement of the values of inductors (L) and capacitors (C) at radio frequencies differs somewhat from the same measurements at low frequencies. Although similarities exist, the RF measurement is a bit more complicated. One of the reasons for this situation is that stray or "distributed" inductance and capacitance values of the test set-up will affect the results. Another reason is that capacitors and inductors are not ideal components, but rather all capacitors have some inductance, and all inductors have capacitance. In this chapter we will take a look at several methods for making such measurements.

VSWR METHOD

When a load impedance (R + jX) is connected across an RF source, the maximum power transfer occurs when the load impedance (Z_L) and source (Z_S) impedances are equal ($Z_L = Z_S$). If these impedances are not equal, then the *voltage standing wave ratio* (VSWR) will indicate the degree of mismatch. We can use this phenomenon to measure values of inductance and capacitance using the scheme shown in Figure 13.1A. The instrumentation required includes a signal generator or other signal source, and a VSWR meter or VSWR analyzer.

Some VSWR instruments require a transmitter for excitation, but others will accept the lower signal levels that can be produced by a signal generator. An alternative device is the SWR analyzer type of instrument. It contains the signal generator and VSWR meter, along with a frequency counter to be sure of the actual test frequency. Whatever signal source is used, however, it must have a variable output frequency. Further, the frequency readout must be accurate (the accuracy of the method depends on knowing the actual frequency).

The load impedance inside the shielded enclosure consists of a noninductive resistor (R1) that has a resistance equal to the desired system impedance resistive component (50 ohms in most RF applications, and 75 ohms in television and video).

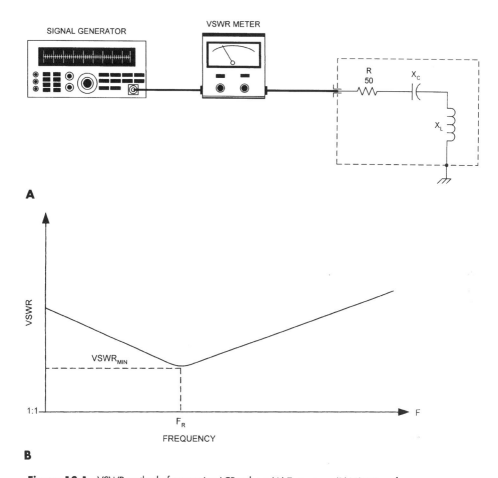

Figure 13.1 VSWR method of measuring LCR values: (A) Test set-up; (B) VSWR-vs.-frequency curve.

An inductive reactance (X_L) and a capacitive reactance (X_C) are connected in series with the load. The circuit containing a resistor, capacitor, and inductor simulates an antenna feedpoint impedance. The overall impedance is:

$$Z_L = \sqrt{R^2 + (X_L - X_C)^2}$$ [13.1]

Note the reactive portion of Equation 13.1. When the condition $|X_L| = |X_C|$ exists, the series network is at resonance, and VSWR is minimum (Figure 13.1B). This gives us a means for measuring the values of the capacitor or inductor, provided that the other is known. That is, if you want to measure a capacitance, then use an inductor of known value. Alternatively, if you want to know the value of an unknown inductor, use a capacitor of known value.

Using the test set-up in Figure 13.1A, adjust the frequency of the signal source to produce minimum VSWR.

1. For finding an inductance from a known capacitance:

$$L_{\mu H} = \frac{10^{12}}{4\pi^2 f^2 C_{PF}} \qquad [13.2]$$

Where:

$L_{\mu H}$ = inductance in microhenrys (μH)
C_{pF} is the capacitance in picofarads (pF)
f is the frequency in hertz (Hz)

2. For finding a capacitance from a known inductance:

$$C_{PF} = \frac{10^{12}}{4\pi^2 f^2 L_{\mu H}} \qquad [13.3]$$

The accuracy of this approach depends on how accurately the frequency and the known reactance are known, and how accurately the minimum VSWR frequency can be found.

VOLTAGE DIVIDER METHOD

A resistive voltage divider is shown in Figure 13.2A. This circuit consists of two resistors (R1 and R2) in series across a voltage source V. The voltage drops across R1 and R2 are V1 and V2, respectively. We know that either voltage drop is found from:

$$V_X = \frac{V R_X}{R1 + R2} \qquad [13.4]$$

Where:

V_X is V1 and R_X is R1 or, V_X is V2 and R_X is R2, depending on which voltage drop is being measured.

We can use the voltage divider concept to find either inductance or capacitance by replacing R2 with the unknown reactance. Consider first the inductive case. In Figure 13.2B resistor R2 has been replaced by an inductor (L). The resistor

R1 is the inductor series resistance. If we measure the voltage drop across R1 (i.e., "E" in Figure 13.2B), then we can calculate the inductance from:

$$L = \frac{R}{2\pi f} \times \sqrt{\left(\frac{V}{E}\right)^2 - \left(1 + \frac{R_S}{R1}\right)^2} \qquad [13.5]$$

As can be noted in Equation 13.5, if R1 >> R_S, then the quotient $R_S/R1$ becomes negligible. In capacitors the series resistance is typically too small to be of consequence. We can replace L in the model of Figure 13.2B with a capacitor, and again measuring voltage E. The value of the capacitor will be:

$$C = \frac{2\pi f \times 10^6}{R \times \sqrt{\left(\frac{V}{E}\right)^2 - 1}} \qquad [13.6]$$

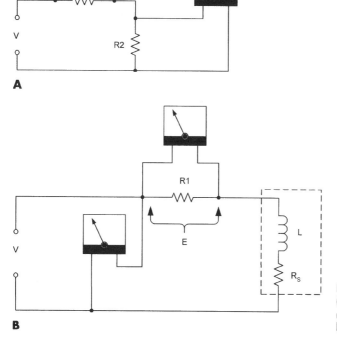

Figure 13.2
(A) Resistive voltage divider network; (B) One resistor replaced by an inductive reactance.

The value of resistance selected for R1 should be approximately the same order of magnitude as the expected reactance of the capacitor or inductor being measured. For example, if you expect the reactance to be, say, between 1K and 10K at some frequency, then select a resistance for R1 in this same range. This will keep the voltage values manageable.

SIGNAL GENERATOR METHOD

If the frequency of a signal generator is accurately known, then we can use a known inductance to find an unknown capacitance, or a known capacitance to find an unknown inductance. Figure 13.3 shows the test set-up for this option. The known and unknown components (L and C) are connected together inside a shielded enclosure. The parallel-tuned circuit is lightly coupled to the signal source and the display through very low value capacitors (C1 and C2). The rule is that the reactances of C1 and C2 should be very high compared with the reactances of L and C.

The signal generator is equipped with a 6-dB resistive attenuator in order to keep its output impedance stable. The output indicators should be any instrument that will read the RF voltage at the frequency of resonance. For example, you could use either an RF voltmeter or an oscilloscope.

The procedure requires tuning the frequency of the signal source to provide a peak output voltage reading on the voltmeter or 'scope. If the value of one of the components (L or C) is known, then the value of the other can be calculated using Equation 13.2 or 13.3, as appropriate.

Alternate forms of coupling are shown in Figure 13.4. In either case, the idea is to isolate the instruments from the L and C elements. In Figure 13.4A, the isolation is provided by a pair of high value (10K to 1 Meg) resistors, R1 and R2. In Figure 13.4B the coupling and isolation is provided by a one- or two-turn-link winding over the inductor. The links and the main inductor are lightly coupled to each other.

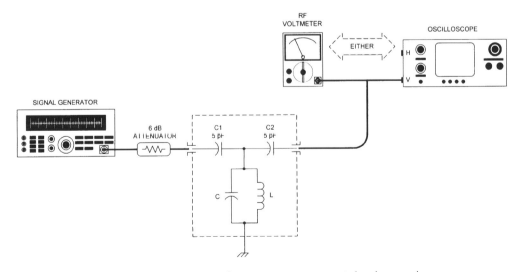

Figure 13.3 Measuring an L-C tuned circuit using capacitors to isolate the network.

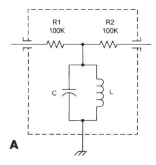
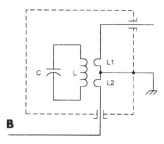

Figure 13.4 Additional coupling isolation methods for L-C circuit measurements: (A) resistive; (B) transformer link.

FREQUENCY-SHIFTED OSCILLATOR METHOD

The frequency of a variable-frequency oscillator (VFO) is set by the combined action of an inductor and a capacitor. We know that a change in either capacitance or inductance produces a frequency change equal to the square of the component ratio. For example, for an inductance change:

$$L2 = L1 \times \left[\left(\frac{F1}{F2} \right)^2 - 1 \right]$$
[13.7]

Where:

L1 is the original inductance
L2 is the new inductance
F1 is the original frequency
F2 is the new frequency

From this equation we can construct an inductance meter such as Figure 13.5. This circuit is a Clapp oscillator designed to oscillate in the high frequency (H.F.) range up to about 12 MHz. The components L1, C2, and C2 are selected to resonate at some frequency. Inductor L1 should be of the same order of magnitude as L1. The idea is to connect the unknown inductor across the test fixture terminals. Switch S1 is set to position "b" and the frequency (F1) is measured on a digital frequency counter. The switch is then set to position "a" in order to put the unknown inductance (L2) in series with the known inductance (L1). The oscillator output frequency will shift to F2. When we know L1, F1, and F2 we can apply Equation 13.7 to calculate L2.

If we need to find a capacitance, then modify the circuit to permit a capacitance to be switched into the circuit across C1 instead of an inductance as shown in Figure 13.5. Replace the "L" terms in Equation 13.7 with the corresponding "C" terms.

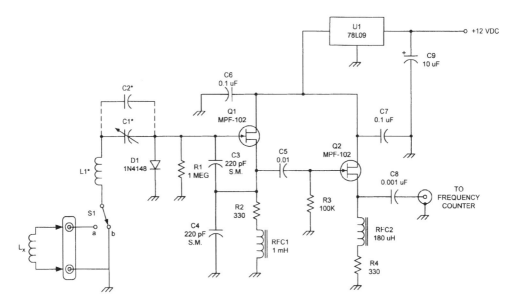

Figure 13.5 Oscillator shift method for making inductance measurements.

Using RF Bridges

Most RF bridges are based on the DC *Wheatstone bridge* circuit (Figure 13.6). In use since 1843, the Wheatstone bridge has formed the basis for many different measurement instruments. The *null condition* of the Wheatstone bridge exists when the voltage drop of R1/R2 is equal to the voltage drop of R3/R4. When the condition R1/R2 = R3/R4 is true, then the voltmeter (M1) will read zero. The basic measurement scheme is to know the values of three of the resistors, and use them to measure the value of the fourth. For example, one common scheme is to connect the unknown resistor in place of R4, make R1 and R3 fixed resistors of known value, and R2 is a calibrated potentiometer marked in ohms. By adjusting R2 for the null condition, and then reading its value, we can use the ratio (R2 × R3)/R1 = R4.

The Wheatstone bridge works well for finding unknown resistances from DC to some relatively low RF frequencies, but to measure L and C values at higher frequencies we need to modify the bridges. Three basic versions are used: Maxwell's bridge (Figure 13.7), Hay's bridge (Figure 13.8), and Schering's bridge (Figure 13.9).

Maxwell Bridge

The Maxwell bridge is shown in Figure 13.7. The null condition for this bridge occurs when:

$$L1 = R2 \times R3 \times C1 \qquad [13.8]$$

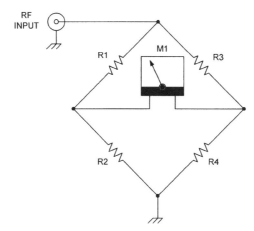

Figure 13.6
Simple Wheatstone bridge.

and,

$$R4 = \frac{R2 \times R3}{R1} \qquad [13.9]$$

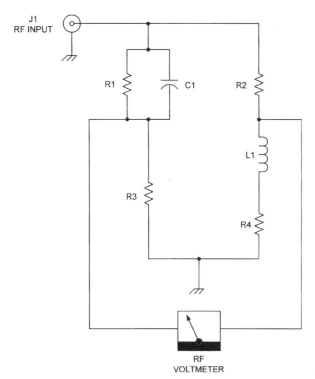

Figure 13.7
Maxwell's bridge.

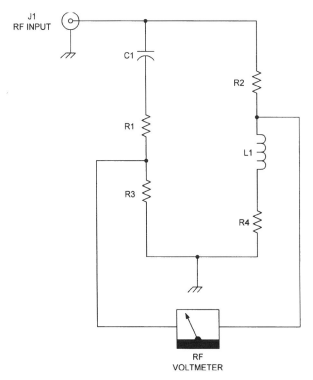

Figure 13.8
Hay's bridge.

RF
VOLTMETER

The Maxwell bridge is often used to measure unknown values of inductance (e.g., L1) because the balance equations are totally independent of frequency. The bridge is also not too sensitive to resistive losses in the inductor (a failing of some other methods). Additionally, it is much easier to obtain calibrated standard capacitors for C1 than it is to obtain standard inductors for L1. As a result, one of the principal uses of this bridge is inductance measurement.

Maxwell bridge circuits are often used in measurement instruments called *Q-meters*, which measure the quality factor (Q) of inductors. The equation for Q is, however, frequency sensitive:

$$Q = 2 \times \pi \times F \times R1 \times C1 \qquad [13.10]$$

Where:

F is in Hertz, R1 in ohms, and C1 in farads

Hay Bridge

The Hay bridge (Figure 13.8) is physically similar to the Maxwell bridge, except that the R1/C1 combination is connected in series rather than parallel. The Hay

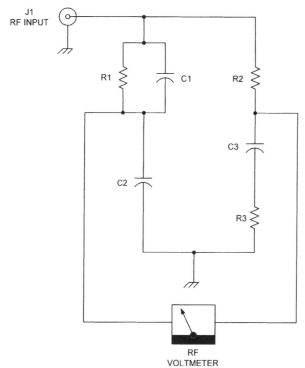

Figure 13.9
Schering's bridge.

RF
VOLTMETER

bridge is, unlike the Maxwell bridge, frequency sensitive. The balance equations for the null condition are also a little more complex:

$$L1 = \frac{R2 \times R3 \times C1}{1 + \left[\dfrac{1}{Q}\right]^2} \qquad [13.11]$$

$$R4 = \left[\frac{R2 \times R3}{R1}\right] \times \left[\frac{1}{Q^2 + 1}\right] \qquad [13.12]$$

Where:

$$Q = \frac{1}{\omega \times R1 \times C1} \qquad [13.13]$$

The Hay bridge is used for measuring inductances with high Q figures, while the Maxwell bridge is best with inductors that have a low Q value.

Note: A frequency independent version of Equation 13.11 is possible when Q is very large (i.e., > 100):

$$L1 = R2 \times R3 \times C1 \qquad [13.14]$$

Schering Bridge

The Schering bridge circuit is shown in Figure 13.9. The balance equation for the null condition is:

$$C3 = \frac{C2 \times R1}{R2} \qquad [13.15]$$

$$R3 = \frac{C2 \times R1}{R2} \qquad [13.16]$$

The Schering bridge is used primarily for finding the capacitance and the power factor of capacitors. In the latter applications, no actual R3 is connected into the circuit, making the series resistance of the capacitor being tested (e.g., C3) the only resistance in that arm of the bridge. The capacitor's Q factor is found from:

$$Q_{C3} = \frac{1}{\omega \times R1 \times C1} \qquad [13.17]$$

FINDING PARASITIC CAPACITANCES AND INDUCTANCES

Capacitors and inductors are not ideal components. A capacitor will have a certain amount of series inductance (called "parasitic inductance"). This inductance is created by the conductors in the capacitor, especially the leads. In older forms of capacitor, such as the wax paper dielectric devices used prior to about 1960, the series inductance was very large. Because the inductance is in series with the capacitance of the capacitor, it forms a series-resonant circuit.

Figure 13.10 shows a test set-up for finding the series-resonant frequency. A *tracking generator* is a special form of sweep generator that is synchronized to the frequency sweep of a spectrum analyzer. It is used with spectrum analyzers in order to perform stimulus-response measurements such as those shown in Figure 13.10.

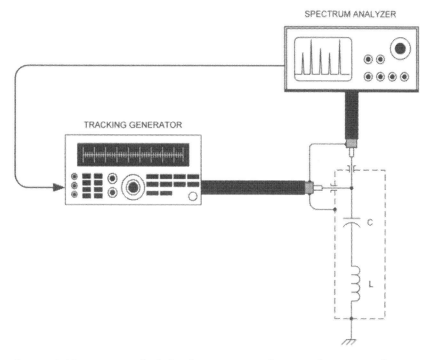

Figure 13.10 A test set-up for finding the series resonant frequency of an L-C network using a tracking generator and spectrum analyzer.

The nature of a series-resonant circuit is to present a low impedance at the resonant frequency, and a high impedance at all frequencies removed from resonance. In this case (Figure 13.10), that impedance is across the signal line. The display on the spectrum analyzer will show a pronounced, sharp dip at the frequency where the capacitance and the parasitic inductance are resonant.

The value of the parasitic series inductance is:

$$L = \frac{1}{2^2 \, \pi^2 f^2 \, C} \qquad\qquad [13.18]$$

Inductors are also less than ideal. The adjacent turns of wire form small capacitors, which when summed up can result in a relatively large capacitance value. Figure 13.11 shows a method for measuring the parallel capacitance of an inductor.

Because the capacitance is in parallel with the inductance, it forms a parallel resonant circuit. These circuits will produce an impedance that is very high at the resonant frequency, and very low at frequencies removed from resonance. In Figure 13.11 the inductor and its parasitic parallel capacitance are in series with the

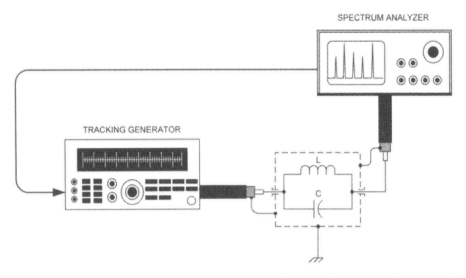

Figure 13.11 A test set-up for finding the parallel resonant frequency of an L-C network using a tracking generator and spectrum analyzer.

signal line, so they will (like the other circuit) produce a pronounced dip in the signal at the resonant frequency. The value of the parasitic inductance is:

$$C = \frac{1}{2^2\,\pi^2 f^2\,L} \qquad [13.19]$$

There are other forms of bridge, and other methods, for measuring L and C elements in RF circuits, but the ones discussed above are very practical, especially in the absence of specialist instrumentation.

CHAPTER FOURTEEN

Time-Domain Reflectometry

Transmission line difficulties can be a notoriously difficult to solve, especially when the load end is not easily accessible (e.g., at the top of an antenna). Although there are a number of different methods available, this chapter discusses time-domain reflectometry (TDR). Although a professional time-domain reflectometer is an expensive piece of equipment, some TDR methods can be used by anyone who has access to an oscilloscope and a pulse generator. Although the professional TDR instrument will be more accurate and have other features, the simple oscilloscope is simply implemented by anyone.

THE BASIS OF TDR

Time-domain reflectometry works on the principle that waves propagating along an unmatched transmission line reflect from the load back towards the source (Figure 14.1). The waveform seen at any given point along the line is the algebraic sum of the forward or "incident" waveform (V_i) and reflected waveform (V_r). When they combine, the incident and reflected waveforms produce an analyzable pattern (Figure 14.2). In TDR measurements, we look at the waveform at the input end of the transmission line system, so we do not need to have access to the load end.

Figure 14.3 shows the basic set-up for our impromptu TDR. A pulse generator, or other source of 1-MHz square waves, is applied simultaneously to the vertical input of an oscilloscope and the input end of the transmission line. This neat little trick is accomplished with an ordinary coaxial tee connector, either BNC or UHF (PL-259/SO-239), depending on your own situation.

The Pulse Source

Almost any source of square waves in the vicinity of 1 MHz can be used for the pulse generator. If you have one of those function generators that output pulses to

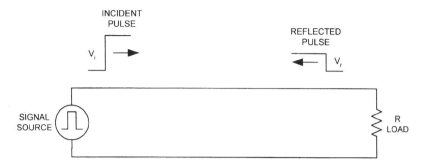

Figure 14.1 Basis for time-domain reflectometry is the algebraic adding of forward and reflected signals at transmission line input end.

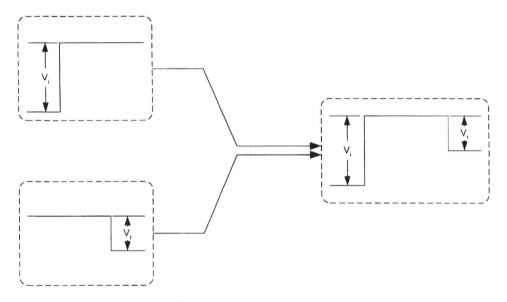

Figure 14.2 Forward and reflected signals combine to form a composite voltage.

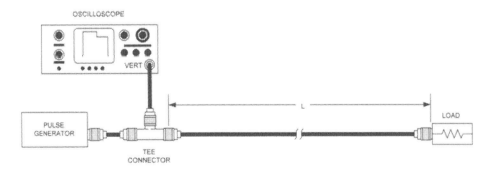

Figure 14.3 Time-domain reflectometry (TDR) test set-up.

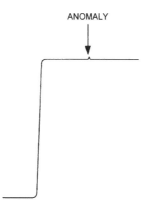

ANOMALY

Figure 14.4
Anomaly due to either coax
connectors or defect in
transmission line.

1 MHz or more, then use it. Be careful, however, if the signal generator output impedance is 600 ohms (as some are). For those signal generators, you might want to fashion a 600- to 50-ohms transformer, or a simple resistor pad. Alternatively, you can build your own pulse source. The signal generator output waveform is a square wave with a period of about 1 μs and a fast rise time.

Figure 14.4 is provided for comparison when considering the following waveforms. It represents the oscilloscope waveform with the coaxial cable perfectly matched. In the waveforms to follow we will see what this pulse looks like when a reflected pulse combines with it after returning down the transmission line. There is an anomaly seen in the upper porch of the pulse waveform. This is probably due to reflection caused by the coaxial connectors. Because the anomaly is approximately at the middle of the waveform, the factor that caused it had to be at or close to the end of the line.

RESULTS

Figure 14.5 shows the results of two different situations. In Figure 14.5A the load resistance is higher than the characteristic impedance of the line ($Z_L > Z_O$), and in Figure 14.5B the opposite is true ($Z_L < Z_O$).

The propagation time along the line can be found by measuring the segment marked T in Figures 14.5A and 14.5B. The line length can be determined by:

$$Length = \frac{V_O T}{2}$$

Where:

V_O is the velocity of propagation
T is time in seconds

The velocity of propagation is compared with the speed of light ($c = 2.99 \times 10^8$ meters/second).

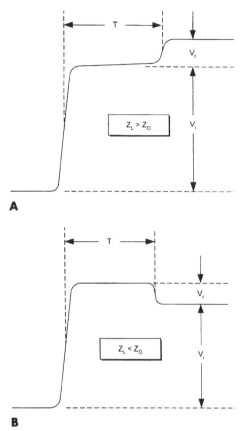

Figure 14.5
Situation when (A) $Z_L > Z_O$, and (B) $Z_L < Z_O$.

A

B

Transmission line *velocity factor* (VF) is the fraction of *c* that waves propagate in the line. The value of VF can be found in the cable manufacturers' data sheets, but in general the following values are generally close:

1. Polyethylene dieletric   0.66
2. Teflon® dielectric   0.70
3. Polyfoam dielectric   0.80

The value of time T can be determined from the grid on the face of the oscilloscope screen and the timebase setting of the 'scope.

Cable Length Measurement

The value of T is 4.8 divisions on an oscilloscope with the time-base control set to 50 nanoseconds per division (5×10^{-8} s/div). Find the length of the cable if polyfoam dielectric cable (VF = 0.80) is being tested.

$$Length = \frac{(0.80 \times 2.99 \times 10^8 \ m/s) \ (5 \times 10^{-8} \ s/cm)(4.8 \ cm)}{2}$$

$$Length = \frac{57.41 \ m}{2} = 28.705 \ meters$$

The accuracy of the length determined by this method depends on the accuracy of the time-base of the oscilloscope.

Velocity Factor Measurement

The velocity factor (VF) is folded into the propagation velocity, V_P, so this same method can be used to determine the velocity factor of a line of known length (L).

$$VF = \frac{2\,L}{2\,c\,T}$$

Assume $T = 3.3 \times 10^{-7}$ second and the cable length is 65 meters.

$$VF = \frac{(2)(65\,m)}{(2)(2.99 \times 10^8\,m/s)(3.3 \times 10^{-7}\,s)} = \frac{130\,m}{1.973 \times 10^2\,m} = 0.659$$

SWR Measurement

The TDR method of this chapter provides the incident and reflected voltages, so we can use them to measure the VSWR of the transmission line and load:

$$VSWR = \frac{V_i + V_r}{V_i - V_r}$$

Suppose $V_i = 4.2$ volts, and $V_r = 1.4$ volts, what is the VSWR?

$$VSWR = \frac{4.2 + 1.4}{4.2 - 1.4} = \frac{5.6}{2.8} = 2:1$$

Once we know the VSWR we can also:

1. Calculate load impedance by knowing the characteristic impedance of the coaxial cable, or
2. Calculate the characteristic impedance of the coaxial cable by knowing the load impedance.

These latter capabilities are due to the fact that VSWR is determined by the ratio of the two impedances.
For $Z_L > Z_O$:

$$VSWR = \frac{Z_L}{Z_O}$$

For $Z_L < Z_O$:

$$VSWR = \frac{Z_O}{Z_L}$$

Select the correct version of the equation for the situation, and then solve for the unknown impedance.

BIBLIOGRAPHY

Ballo, David. (1997). *Network Analyzer Basics.* Back to Basics Seminar. Englewood, CO: Hewlett-Packard.

Bates, Herbert. "Stray Capacitance Affects Inductance Measurements." *QST.* Sept 1988: 44. Newington, CT.

Belrose, John S. "RX Noise Bridges." *QST.* May 1988. Newington, CT.

Booth, Lionel. "Determination of Inductance or Capacitance Using SWR Measurements." *QST.* Oct 1994: 75. Newington, CT.

Brown, Christie. (1997). *Spectrum Analyzer Basics.* Back to Basics Seminar. Englewood, CO: Hewlett-Packard.

Camillo, Charles. "A Reevaluation of the Caron RF Impedance Bridge." *QST.* Sept 1994. Newington, CT. (See correction in *QST,* Nov 1994, p. 88).

Campbell, N.R. (1957). *Foundations of Science.* New York: Dover.

Carr, Joseph J. (1999). "The RF Hybrid Coupler." *Electronics World/Wireless World.* Surrey, UK: Sutton. 41–43.

Carr, Joseph J. (1998). *Practical Antenna Handbook.* 3rd Ed. New York: McGraw-Hill.

Carr, Joseph J. (1997). *Microwave and Wireless Communications Technology.* Boston: Newnes.

Carr, Joseph J. (1996). *Secrets of RF Circuit Design.* 2nd Ed. New York: McGraw-Hill.

Dye, Norm, and Helge Granberg. (1993). *Radio Frequency Transistors.* Boston: Butterworth–Heinemann.

Grebenkemper, John. "Improving and Using the R-X Noise Bridges." *QST.* August 1989: 27–32. Newington, CT.

Hagen, Jon B. (1996). *Radio-Frequency Electronics: Circuits and Applications.* Cambridge, UK: Cambridge University Press.

Hardy, James. (1979). *High Frequency Circuit Design.* Reston, VA: Reston/Prentice-Hall.

Herceg, E.E. (1972). *Handbook of Measurement and Control.* Pennsauken, NJ: Schaevitz Engineering.

Hewlett-Packard Application Note 1287–1. (1997). *Understanding the Fundamental Principles of Vector Network Analysis.* Englewood, CO.

Hewlett-Packard Application Note 1287–2. (1997). *Exploring the Architectures of Network Analyzers.* Englewood, CO.

Hewlett-Packard Application Note 1287–3. (1997). *Applying Error Correction to Network Analyzer Measurements.* Englewood, CO.

Hewlett-Packard Application Note 1287–4. (1997). *Network Analyzer Measurements: Filter and Amplifier Examples.* Englewood, CO.

Hewlett-Packard Application Note 1288–4. (1997). *How to Characterize CATV Amplifiers Effectively.* Englewood, CO.

Kinley, R. Harold. (1985). *Standard Radio Communications Manual: With Instrumentation and Testing Techniques.* Englewood Cliffs, NJ: Prentice Hall.

Laverghetta, Thomas S. (1984). *Practical Microwaves.* Indianapolis, IN: Howard W. Sams.

Liao, Samuel Y. (1990). *Microwave Devices and Circuits.* Englewood Cliffs, NJ: Prentice Hall.

Mandel, John. (1964). *The Statistical Analysis of Experimental Data.* New York: John Wiley.

Noble, Frank. "Capacitance Measurement with a Dip Meter." *QST.* Dec 1980. Newington, CT.

Phelps, Darin. (1997). *Measurement Solutions for Testing Base-Station Amplifiers.* Santa Rosa, CA: Hewlett-Packard.

Sabin, William E., and Edgar O. Schoenike, editors. (1998). *HF Radio Systems and Circuits.* 2nd Ed. Atlanta: Noble Publishing.

Sevick, Jerry. "Simple RF Bridges." *QST.* April 1975. Newington, CT.

Shrader, Robert L. (1975). *Electronic Communication.* 3rd Ed. New York: McGraw-Hill.

Test and Measurement Educator's Corner. *Teacher's Tools.* http://www.tmo.hp.com/

Test and Measurement Educator's Corner. "Network Analyzer Basics: Power Sweep—Gain Compression." http://www.tmo.hp.com/

Test and Measurement Educator's Corner. "Network Analyzer Basics: Power Sweep-AM to PM Conversion. http://www.tmo.hp.com/

Tracy, M., and M. Gruber. (1998). *Test Procedures Manual.* Newington, CT: ARRL.

Vizmuller, Peter. (1995). *RF Design Guide.* Boston/London: Artech House.

Index

16736601R00190

Made in the USA
Lexington, KY
08 August 2012